PURE AND APPLIED BIOGEOGRAPHY

Edited by **Levente Hufnagel**

Pure and Applied Biogeography
http://dx.doi.org/10.5772/65139
Edited by Levente Hufnagel

Contributors
Ryan McManamay, Brenda Pracheil, Natalie Griffiths, Christopher DeRolph, Alina Morquecho-Contreras, Carmen Zepeda Gómez, Hermilo Sánchez-Sánchez, Melinda Pálinkás, Julián A. Velasco, Mamokete N.V. Dingaan, Johann Du Preez, Joel Loitu Meliyo, Kenneth F.G. Masuki, Balthazar M. Msanya, Didas N. Kimaro, Loth S. Mulungu, Levente Hufnagel

Published by InTech
Janeza Trdine 9, 51000 Rijeka, Croatia

Publishing Process Manager Nina Kalinic
Technical Editor SPi Global
Cover InTech Design team

First published January, 2018
Printed in Croatia
Legal deposit, Croatia: National and University Library in Zagreb

Additional hard copies can be obtained from orders@intechopen.com

Pure and Applied Biogeography, Edited by Levente Hufnagel
p. cm.
Print ISBN 978-953-51-3754-2
Online ISBN 978-953-51-3755-9

For EU product safety concerns:
IN TECH d.o.o., Prolaz Marije Krucifikse Kozulić 3, 51000 Rijeka, Croatia,
info@intechopen.com or visit our website at intechopen.com.

Meet the editor

Dr. Levente Hufnagel is an associate professor and senior researcher at Szent István University, Gödöllő, Hungary, in climate change and ecosystem ecology, biogeography, biometrics, and ecological research methodology. He has over 20 years of experience at leading Hungarian academic institutions teaching PhD, MSc, and BSc students from various social and cultural backgrounds, having more than 140 scientific publications (in both aquatic and terrestrial ecological aspects of plants, animals, and microbes, at community as well as population level) and more than 500 independent citations. He is a participant of several big ecological research and development projects, having significant experience in multidisciplinary cooperations (with more than 180 coauthors in different publications). He is the supervisor of several PhD, BSc, and MSc theses and editor in chief of an international scientific journal indexed by Web of Science and Scopus having wide supervising and editing experience. Dr. Hufnagel graduated from Eötvös Lorand University with a master degree in Ecology and Evolutionary Biology and a PhD degree in Hydrobiology; he also graduated from Szent István University with a PhD degree in Agricultural Science.

Contents

Preface

This book—*Pure and Applied Biogeography*—gives a very interesting report and overview about the frontiers of such parts of recent biogeographical research, which plays important roles in solving our most pressing global problems (biodiversity crisis, climate change, water issues, and sustainable agriculture).

Our book consists of three sections: "Introduction", "Pure Biogeography and Global Patterns" and "Applied Biogeography and Regional Issues."

After the introductory chapter, which is about the main branches and aims of biogeography in service of solving global problems, we can find three chapters as parts of the first section.

The first chapter in this section is in close relation with the origin of biodiversity and conservation. The second and third chapters are about the biogeographical aspects of climate change and biodiversity.

In the second section of this book, three applied biogeographical chapters can be found, which are related to agriculture, theoretical background of biological plant protection against herbivores, and regional patterns in ecological biogeography.

I am sure that this book will be very useful for everybody—researchers, professors, students, or others interested in the field—who would like to get an overview about the most practical results and research areas of biogeography nowadays.

Levente Hufnagel, PhD habil
Szent István University,
Gödöllő, Hungary

Section 1

Introduction

Introductory Chapter: The Main Directions and Tasks of Pure and Applied Biogeography in Solving the Global Problems of Our Time

Levente Hufnagel

Additional information is available at the end of the chapter

http://dx.doi.org/10.5772/intechopen.72873

Pure and applied biogeography is the study of the spatial and temporal distribution of taxa, life forms, communities, biomes, and natural or human-influenced ecosystems of our planet in large scales.

Biogeography is one of the oldest life sciences, because already since Alexander von Humboldt's work (1769–1859) it has been a theoretically grounded discipline.

Phytogeography and zoogeography are classical subdisciplines of biogeography that study the spatial and temporal distribution of plants and animals, the flora and fauna. Also important recent subdisciplines are island biogeography [1, 2], phylogeography [3], spatial population genetics [4], paleobiogeography [5], systematic and evolutionary biogeography [6–8], fragmentation, metapopulation and landscape biogeography [9–11], gradient analyses [12], and ecogeography [13].

However, many new theoretical trends have emerged recently [14, 15], a number of new methods are used [16], and biogeography has gained outstanding social significance through the effects of human land use on natural communities [17].

The fastest growing trends in biogeography are those that are closely related to quantitative ecology [18–20]. Quantitative ecology is one of the most important disciplines of our time, which is an indispensable part of ecology, environmental science, agricultural sciences, research methodology, biometrics, and also biogeography. Quantitative ecology is about the application of biomathematics, environmental informatics, and biostatistical methods in life and environmental sciences that relate to data collection, data analysis, modeling, monitoring, evaluation, and data communication tasks of supraindividual organizations. Quantitative approach has great significance in biogeography where large quantities of observation, measurement, experiment, or literature data are available in complex systems, processes, or phenomena.

1. Global problems

Overpopulation is in the center of the global problems of humanity. Overpopulation is the primary cause of many other global problems, which are also influenced by some other minor factors. These global crisis phenomena are closely related.

- Overpopulation is the cause of an increase in the ecological footprint of humanity. This is manifested in the destruction of natural ecosystems, depletion of natural resources, pollution of air, water, and soil [21], an increase in anthropogenic climate change, global biodiversity loss, land-use restructuring [22], and urbanization problems.
- Overpopulation increases density and decreases personal space and per capita resources. The human-ethological consequence of these is the growing aggression that is manifested in crimes, terrorism, revolutions, wars, and so on.
- Overpopulation and urbanization increase the severity of global health problems and increase the risk of pandemics.

There are a lot of synergistic interactions between these phenomena, for example:

- Decreasing natural ecosystems (tropical rain forests, moderate forests, bogs, marshes, seaside habitats, and coral reefs) reduces the biosphere's climate control capacity.
- Growing populations are mining and burning more and more fossil fuels, which increases greenhouse gas emissions and the rate of global warming.
- Global climate change increases demand for adaptability, living costs, poverty, and social aggression and enhances migration.
- Global climate change further undermines the living conditions of natural ecosystems.
- Global climate change creates conditions for outbreaks of new epidemics (through area change of parasitic, vectorial, and pathogenic organisms).
- Reducing biodiversity reduces the biosphere's adaptability (through decreasing functional redundancy).
- The deterioration of people's living conditions increases the number of people in extreme poverty, which further aggravates overpopulation.

These problems are further aggravated by the growing social differences between people, the problems of low schooling of people in poverty, and the lack of political unity of mankind (195 nonaligned nation states).

To address problems, overcome the overcrowding, reduce the per capita ecological footprint, tighten coordination of human activity, reduce social disparities, strengthen social solidarity, environmental protection, nature conservation, and climate protection, improve the education level of humankind, and improve public health, resources may be concentrated on vital scientific research. This way we can develop a sustainable global society in a healthy environment and rich biosphere.

In solving our global problems, biogeographical research plays a key role in four main areas:

- Origin and protection of biodiversity
- Global climate change issues
- Water issues, aquatic ecosystems
- Sustainable agriculture, biological pest control issues

2. Biodiversity

In the maintenance and conservation of our planet's biodiversity, knowledge of current biogeographical patterns [23], Earth-historical changes [24, 25], and speciation processes [26] have outstanding importance. Many areas of biodiversity are still completely undiscovered nowadays [27].

The ecological effects of global climate change also have a significant impact on biodiversity, flora and fauna, through biogeographical patterns [28–31].

In the field of nature conservation, a fundamental change of paradigm became necessary due to climate change. The previously dominant "in situ conservation," which is to preserve existing ecological conditions in present habitats, is not always a realistic goal. Consideration should be given to active "eco-engineering" interventions that support the spatial shifts of natural communities as adaptation options. In this, the biogeographic knowledge and the identification of climate-analogous areas could have great importance. It is obvious that such studies should combine analyses of current and Earth-historical biogeographic patterns with climatic scenario-based predictions.

In the biogeographical research on biodiversity, the following key issues can be identified:

- Tropical rainforests, coral reefs, and other biodiversity hotspots
- Natural and near-natural habitats
- Human-influenced habitats and invasions
- Cross-border phenomena
- Paleobiogeography of biodiversity
- Conservation biogeography
- Biogeography of ecosystem services
- Applied biogeographical research in service of sustainable agriculture and organic farming

3. Climate change

Climate change—in close connection with overpopulation, global species extinction and biodiversity crisis, and social crises—is the biggest challenge for mankind in our history. An adequate and stable global climate system is the most basic living condition of the biosphere, and as a part of it, human society and even all other living conditions are dependent on climate. This climatic system also determines the basic biogeographic patterns of our Earth. The fact and the anthropogenic origin of climate change are supported by evidence of a multitude of research findings and observations. With regard to these two things, there is an unprecedented degree of consensus in the scientific world.

In the background of this phenomenon, there are a lot of causes in strict connection with each other like overpopulation of mankind, the fossil fuel production and burning since the industrial revolution, the rapid eradication of forests and natural wetlands, industrial and transport pollution, unsustainable agricultural practices, and the rapid urbanization. Human activity is causing the greatest damage and dangers by breaking down the healthy functioning and regulatory capacity of the Earth's biosphere, eradicating primeval forests and rainforests, draining bogs and swamps, and reducing forest cover. The importance of grazing is also well known in the diversity of nature or near-nature grasslands [32].

There are a number of well-founded and less well-founded opinions about climate change today, but there are some important points that are beyond dispute:

1. The climate determines the living conditions of ecosystems and, at the same time, human society, which are also reflected in large-scale biogeographical patterns.
2. Past climate changes have always had significant ecological impacts, mass extinction of species, new species becoming dominant, and a fundamental change in landscape.
3. The history of the Earth shows a great deal of climate changes, so it must be our basic attitude that, because climate is variable, it changes and the biogeographical and social consequences can be very serious if we are not prepared for them.
4. A better understanding of the relationship between climate and biosphere should be seen as one of our most important research tasks because the existence and prosperity of humanity depend on it.
5. Today, it has become an undeniable fact that the collective activity of mankind is a decisive field-altering, environment-changing factor of our planet, which, besides everything else, also has a significant impact on climate.
6. In our time, global crisis phenomena (biodiversity crisis, raw material and energy problems, consequences of overpopulation and climate change) interact with one another in a synergistic way.

Natural ecosystems provide carbon dioxide binding and storing functions, feedback processes, and in a number of direct and indirect ways regulate the climate in a biogeographical scale, and provide retention and dispensing of leaking residues, as well as the development of favorable micro- and mesoclimate. More recently, it has come to light that seaweeds emit dimethyl sulfide [33] and some pine trees emit terpenes from themselves [34, 47] as a result of rising temperature and carbon dioxide level; these can effectively promote cloud formation and so cool our planet. In addition, there are probably many ecological phenomena and processes so far unknown to science that have a prominent role in the biological regulation of climate [35] and which could also help the survival of the civilization of mankind if we understood and recognized them in time.

In the biogeographical research on global climate change, the following key issues can be identified:

- Climate control potential of different biogeographical units
- The impact of climate change on different biogeographical units
- The role of climate in spatial and temporal biogeographical boundaries and ecological collapses
- Range and time shifts in biogeographical units
- Paleobiogeographical aspects of Earth-historical climate patterns

4. Water and aquatic habitats

Aquatic ecosystems were, for a long time, a neglected area of biogeography, though more than 70% of our planet is covered with water. Water is the foundation of all earthly life, plays a key role in climate change and spatial-temporal patterns of biodiversity, and decisively defines human activity and the possibilities of agriculture.

Climate change and biodiversity crisis issues have a particularly large impact on water, aquatic ecosystems and wet habitats such as oceanic biotopes [36], surface freshwater [37, 38], or groundwater [39]. The state of river water vegetation also has a fundamental effect on animal communities [40].

In the biogeographical research on hydrobiology, the following key issues can be identified:

- Biogeography of oceans and marine habitats
- Biogeography of freshwater habitats and wetlands
- Applied ecogeography of water resources and soil types
- Dynamic biogeography, invasions, and mix-up of communities

5. Sustainable agriculture

The creation of a sustainable and adaptable agriculture is not merely a technological issue, but it is necessary to rethink the whole natural and socioeconomic system related to agriculture from biological, geographical, and human ecological approaches [41].

Sustainable agriculture means a production system where

- The productivity of the agricultural area does not decrease.
- The production does not lead to an increase in the environmental load.
- Soil, air, and natural waters do not get polluted (exceeding their degree of ecological self-purification).
- Soil does not erode.
- Greenhouse gases do not exceed the amount of absorbed quantity during production.
- The quantity and quality (nutritional value and safety) of products produced satisfy social needs.
- The living standards of the family of farmers are ensured.
- Nonagricultural habitats and global biodiversity are also ensured.

The presently dominant conventional agriculture is obviously unable to meet these conditions as agriculture [42]

- is a major source of anthropogenic climate change,
- is a major source of environmental pollution, while
- the productivity of agricultural lands can only be sustained through higher and higher external material and energy inputs (i.e., unsustainable in itself).

For the development of sustainable agricultural systems, only the "high technology" of agriculture is capable: the organic farming sensu lato. This includes organic farming, permaculture, agroforestry, and biodynamic farming.

The productivity, ecological efficiency, and cost-effectiveness of organic farming would be very high (much better than in conventional production) if all of these indicators were to be applied to the unit's environmental use and environmental load.

The global human population boom and at the same time environmental pollution, damaging nature, and land use pose new challenges to agriculture in the field of sustainability and especially organic farming [43]. In this context, the biogeographical researches affect the cultivated plants, weeds, animal pests and pathogens, and their natural enemies [44], as well as the traditional ecological knowledge [45].

In the applied biogeographical research on agriculture, the following key issues can be identified:

- Anthropogenic changes in biogeography, mix-up of flora and fauna [46]
- Biogeography of grown plants, domestic animals, and their potential genetic resources

- Biogeography of weeds, pests and pathogens, and their natural enemies
- Biogeography of soils and soil biological communities
- Ecogeography of climatic patterns in a changing world
- Biogeographical aspects of agroforestry, organic farming, and sustainable agriculture
- Ethnogeographical aspects of traditional ecological knowledge

Ecological and biogeographical research of natural, near-natural, and human-influenced ecosystems has strategic importance in the struggle for survival of mankind and the chance to create a sustainable society.

Author details

Levente Hufnagel

Address all correspondence to: leventehufnagel@gmail.com

Laboratory of Biometrics and Quantitative Ecology, Faculty of Agricultural and Environmental Science, Institute of Crop Production, Szent István University, Gödöllő, Hungary

References

[1] Maria Fernandez-Palacios J. Island shaped by sea-level shifts. Nature. 2016;**532**(7597):42-43

[2] Patino J, Whittaker RJ, Borges PAV. A roadmap for island biology: 50 fundamental questions after 50 years of the theory of island biogeography. Journal of Biogeography. 2017;**44**(5):963-983

[3] Nieto Feliner G. Patterns and processes in plant phylogeography in the Mediterranean Basin. A review. Perspectives in Plant Ecology Evolution and Systematics. 2014;**16**(5): 265-278

[4] Hohn M, Gugerli F, Abran P, Bisztray G, Buonamici A, Cseke K, Hufnagel L, Quintela-Sabaris C, Sebastiani F, Vendramin GG. Variation in the chloroplast DNA of Swiss stone pine (*Pinus cembra* L.) reflects contrasting post-glacial history of populations from the Carpathians and the Alps. Journal of Biogeography. 2009;**36**(9):1798-1806

[5] Seoane FD, Roig Juenent S, Cerdeno E. Phylogeny and paleobiogeography of Hegetotheriidae. Journal of Vertebrate Paleontology. 2017;**37**(1):e1278547

[6] Fattorini S. A history of chorological categories. History and Philosophy of the Life Science. 2016;**38**(3):UNSP 12

[7] Karaveliogullari FA, Celik S, Isik G. The altitudinal, climatical and phonological classification of Verbascum L. species from different phytogeographical regions of Turkey. Applied Ecology and Environmental Research. 2016;**14**(4):15-27

[8] Halffter G, Morrone JJ. An analytical review of Halffter's Mexican transition zone, and its relevance for evolutionary biogeography, ecology and biogeographical regionalization. Zootaxa. 2017;**4226**(1):1-46

[9] Panahandeh M, Yavari AR, Salehi E, Malekmohammadi B. Analysis of habitat changes of Caspian red deer (Cervus eleaphus maral GRAY, 1850) based on temporal changes of effective area and functional connectivity, case study of Lisar protected area. Applied Ecology and Environmental Research. 2016;**14**(4):71-83

[10] Szanyi S, Nagy A, Varga Z. Butterfly assemblages in fragmented meadow habitats of the pre-carpathian lowland (Bereg Plain, SW Ukraine). Applied Ecology and Environmental Research. 2015;**13**(3):615-626

[11] Öztürk M, Bolat İ, Gökyer E, Kara Ö. Growth gradients of multi-aged pure oriental beech stands along the altitudinal gradients within a mesoscale watershed landscape. Applied Ecology and Environmental Research. 2016;**14**(4):101-119

[12] Dyakov NR. Spatial distribution of dominant plant species according to local environmental gradients. Applied Ecology and Environmental Research. 2016;**14**(1):327-347

[13] Kantar MB, Sosa CC, Khoury CK. Ecogeography and utility to plant breeding of the crop wild relatives of sunflower (*Helianthus annuus* L.). Frontiers in Plant Science. 2015;**6**:841

[14] Albert JS, Antonelli A. Society for the study of systematic biology symposium: Frontiers in parametric biogeography. Systematic Biology. 2017;**66**(2):125-129

[15] Lomolino MV, Pijanowski BC, Gasc A. The silence of biogeography. Journal of Biogeography. 2017;**42**(7):1187-1196

[16] Erdős L, Bátori Z, Tölgyesi C, Körmöczi L. The moving split window (MSW) analysis in vegetation science – An overview. Applied Ecology and Environmental Research. 2014;**12**(3):787-805

[17] Plexida SG, Sfougaris AI, Papadopoulos NT. The impact of human land use on the composition and richness of ground and dung beetle assemblages. Applied Ecology and Environmental Research. 2014;**12**(3):661-679

[18] Rédei D, Gaál M, Hufnagel L. Spatial and temporal patterns of true bug assemblages extracted with Berlese funnels. Applied Ecology and Environmental Research. 2003; **1**(1-2):115-142

[19] Gergócs V, Garamvölgyi Á, Hufnagel L. Indication strength of coenological similarity patterns based on genus level taxon lists. Applied Ecology and Environmental Research. 2010;**8**(1):63-76

[20] Gergócs V, Hufnagel L. Global pattern of oribatid mites (*Acari*: *Oribatida*) revealed by fractions of beta diversity and multivariate analysis. International Journal of Acarology. 2015;**41**(7):574-583

[21] Reay DS. Fertilizer 'solution' could turn local problem global – Protecting soil and water from pollution may mean releasing more greenhouse gas. Nature. 2004;**427**(6974):485-485

[22] Yanqiong Y, Jia'en Z, Lili C, Ying O, Prem P. Dynamics of ecosystem service values in response to landscape pattern changes from 1992 to 2005 in Guangzhou, Southern China. Applied Ecology and Environmental Research. 2015;**13**(1):21-36

[23] Freitas WK, Magalhaes LMS, Vives LR. Floristic, diversity and spatial distribution of tree species in a dry forest in Southern Brazil. Applied Ecology and Environmental Research. 2017;**15**(1):511-524

[24] Barnosky AD, Hadly EA, Gonzalez P, et al. Merging paleobiology with conservation biology to guide the future of terrestrial ecosystems. Science. 2017;**355**(6325):594, eaah4787. DOI: 10.1126/science.aah4787

[25] Tennant JP, Mannion PD, Upchurch P. Biotic and environmental dynamics through the Late Jurassic-Early Cretaceous transition: Evidence for protracted faunal and ecological turnover. Biological Reviews. 2017;**92**(2):776-814

[26] Schluter D, Pennel MW. Speciation gradients and the distribution of biodiversity. Nature. 2017;**546**(7656):48-55

[27] Andujar C, Arribas P, Vogler AP. Terra incognita of soil biodiversity: Unseen invasions under our feet. Molecular Ecology. 2017;**26**(12):3087-3089

[28] Garamvölgyi Á, Hufnagel L. Impacts of climate change on vegetation distribution No 1 climate change induced vegetation shifts in the Palearktic region. Applied Ecology and Environmental Research. 2013;**11**(1):79-122

[29] Hufnagel L, Garamvölgyi Á. Impacts of climate change on vegetation distribution No 2, climate change induced vegetation shifts in the new world. Applied Ecology and Environmental Research. 2014;**12**(2):255-422

[30] Sangoony H, Vahabi M, Tarkesh M. Range shift of Bromus tomentellus Boiss. As a reaction to climate change in central Zabros. Iran Applied Ecology and Environmental Research. 2016;**14**(4):85-100

[31] Sakalli A. Impacts of climate change on net primary production: A modelling study at pan-european scale. Applied Ecology and Environmental Research. 2017;**15**(1):1-15

[32] Dingaan MNV, Walker S, Tsubo M, Newby T. Influence of grazing on plant diversity-productivity relationship in semi-arid grassland of South Africa. Applied Ecology and Environmental Research. 2016;**14**(4):1-13

[33] Hussherr R, Levasseur M, Lizotte M. Impact of ocean acidification on Arctic phytoplankton blooms and dimethyl sulfide concentration under simulated ice-free and under-ice conditions. Biogeosciences. 2017;**14**(9):240-2427

[34] Kivimaenpaa M, Ghimire RP, Sutinen S. Increases in volatile organic compound emissions of Scots pine in response to elevated ozone and warming are modified by herbivory and soil nitrogen availability. European Journal of Forest Research. 2016;**135**(2):343-360

[35] Tollefson J. The hostile ocean that slowed climate change. Nature. 2016;**539**(7629):346-348

[36] Hernando-Morales V, Ameneiro J, Teira E. Water mass mixing shapes bacterial biogeography in a highly hydrodynamic region of the Southern Ocean. Environmental Microbiology. 2017;**19**(3):1017-1029

[37] O'Reilly CM, Alin SR, Plisnier PD, Cohen AS, McKee BA. Climate change decreases aquatic ecosystem productivity of Lake Tanganyika, Africa. Nature. 2003;**424**(6950):766-768

[38] Liao J, Cao X, Wang J. Similar community assembly mechanisms underlie similar biogeography of rare and abundant bacteria in lakes on Yungui Plateau, China. Limnology and Oceanography. 2017;**62**(2):723-735

[39] Kambale JB, Singh DK, Sarangi A. Impact of climate change on groundwater recharge in a semi-arid region of Northern India. Applied Ecology and Environmental Research. 2017;**15**(1):335-362

[40] Vieira TB, Dias-Silva K, Pacifico ES. Effects of riparian vegetation integrity on fish and Heteroptera communities. Applied Ecology and Environmental Research. 2015; **13**(1):53-65

[41] Zhang X, Davidson EA, Mauzerall DL, Searchinger TD, Dumas P, Shen Y. Managing nitrogen for sustainable development. Nature. 2015;**528**(7580):51-59

[42] Garnett T, Appleby MC, Balmford A, et al. Sustainable intensification in agriculture: Premises and policies. Science. 2013;**341**(6141):33-34

[43] Benkovic-Lacic T, Brmez M, Pribetic D. Biological diversity of Nematode communities in conventional and organic olive farming. Applied Ecology and Environmental Research. 2016;**14**(2):457-462

[44] Grutters BMC, Roijendijk YOA, Verberk WCEP. Plant traits and plant biogeography control the biotic resistance provided by generalist herbivores. Functional Ecology. 2017; **31**(6):1184-1192

[45] Harisha RP, Padmavathy S, Nagaraja BC. Traditional ecological knowledge (TEK) and its importance in South India: Perspective from local communities. Applied Ecology and Environmental Research. 2016;**14**(1):311-326

[46] Capinha C, Essl F, Seebens H, Moser D, Pereira HM. The dispersal of alien species redefines biogeography in the Anthropocene. Science. 2015;**348**(6240):1248-1251

[47] Materic D, Blankhorn D, Gonzales-Méndez R, Bruhn D, Turner C, Morgan G, Mason N, Gauci V. Monoterpene emission from young scot pine may be influenced by nutrient availability. Applied Ecology and Environmental Research. 2016;**14**(4):667-681

Section 2

Pure Biogeography and Global Patterns

Chapter 2

Are Historical Biogeographical Events Able to Promote Biological Diversification?

Julián A. Velasco

Additional information is available at the end of the chapter

http://dx.doi.org/10.5772/intechopen.69516

Abstract

One of the goals of evolutionary biology is understanding how biological diversification change across spatial and temporal scales. Theoretically, it has been established that external (i.e., dispersals) and internal (i.e., origin of a key innovations) factors can modulate shifts in rates of species diversification. However, the role of historical events as trigger of species diversification rates have not been well understood in empirical studies. I reviewed the literature linking historical biogeographic events and species diversification in many groups. Many of studies conclude that dispersals can be associated with exceptional changes in species diversification rates in insular and mainland areas. I discuss the limitations of some approaches used to discover the link between historical biogeography and macroevolution. I propose some predictions under biogeographic scenarios to gain understanding in how historical events promote biological diversification. I suggest that future studies linking biogeography and macroevolution should incorporate ecologically-relevant traits to discern the mechanisms underlying these historical associations. Although new developments in phylogenetic comparative methods have been done, still is necessary more traditional field-based ecological and evolutionary research. The link between biogeography and diversification still remains narrative and a comprehensive approach is necessary to establish how diversification was triggered by historical events.

Keywords: dispersal, vicariance, ecological opportunity, macroevolution, event-based biogeography

1. Introduction

One the main goals of evolutionary biology is to understand why lineages exhibit differences in species diversification, understood as the difference between speciation and extinction rates in a clade. Three competing hypotheses explain differences in species between clades and geographic regions [10, 51, 69]. The first hypothesis, known as the unbounded hypothesis, proposes that speciation and extinction processes are controlled only by time and diversity is not limited by any ecological process [69]. The second hypothesis, known as the bounded hypothesis, states that diversity reaches a limit imposed by the number of competing coexisting species [51, 52]. And a third recently proposed hypothesis, known as the damped hypothesis [10], states that diversity increases without limit and ecological factors constrain speciation and extinction rates.

The understanding of the historical and ecological circumstances where diversification dynamics occur is key to establish whether diversity is limited or not. Therefore, historical events as dispersals can be associated with shifts in species diversification [40]. However, the mechanisms underlying this association remains elusive for many taxa. For instance, for plant clade Dipsacales, it was found that some increases in species diversification rates were associated with dispersal to new geographic areas [40, 41]. Similarly, in Caribbean Anole lizards, it was found that diversification increases occurred after a lineage colonized an island above a certain island size [30, 50]. These two examples show how diversification dynamics can be linked with historical events and suggest also that macroevolutionary dynamics should be explicitly tested in a historical biogeography context.

With the aim to have a comprehensive understanding of the link between historical biogeography and species diversification, it is necessary to consider which ecological or evolutionary mechanisms were involved in this association. Ecological mechanisms can explain how diversification dynamics take place in a region after a biogeographical event. They also offer a conceptual bridge between historical and ecological biogeography. I assert that historical biogeography events (dispersal and vicariance) can promote biological diversification under some specific ecological and evolutionary mechanisms. For instances, some studies have found that some clades diversified through ecological opportunity [28, 33], biotic interactions [63], and/or climate change [26, 65]). The integration of ecological, phylogenetic, and biogeographic approaches in a comprehensive framework is key to understand why some clades diversify extensively when reach some regions and not others.

Here, I conduct a review of studies evaluating the link between biogeographical events and biological diversification. For biogeographical events, I refer mainly to dispersal and vicariance, which have been considered as competing hypotheses in historical biogeography for many years [45, 60]. Dispersal refers to movements of a lineage to a new region across a geographical (or ecological) barrier. Vicariance refers to the fragmentation of an ancestral geographical range and the emergence of geographical (or ecological) barriers. Here, I first discuss how these two biogeographical events can promote shifts in species diversification (e.g., increases in speciation rates or decreases in extinction rates) or diversification dynamics (from time-dependent to density-dependent cladogenesis). Also, I discuss some limitations of methods used to discover this historical link. I establish some basic specific predictions about the expected evolutionary trajectories of ecologically relevant traits under a dispersal or vicariance scenario and how these predictions might be tested using phylogenetic comparative methods. Finally, I provide some possible avenues for an integration of ecological and evolutionary studies with historical biogeography within an explicit modern phylogenetic comparative framework.

2. Materials and methods

I conducted a literature search in Web of Science using the keywords as *dispersal, vicariance, species diversification,* and *biogeography* during the last 14 years (2002–2016). I selected only those studies that used statistical biogeographical methods to reconstruct ancestral areas in an explicit phylogenetic context (~80 studies). I select those studies where (or at least it was tested) a historical link between past biogeographical events and shifts in diversification dynamics was explicit. Although this review is not exhaustive, my aim here is to discuss some limitations and opportunities of current methods used to study biogeography and diversification.

3. Results and discussion

3.1. Historical associations between biogeography and diversification

Many studies identified that past biogeographic movements (i.e., dispersals) between regions acted as a trigger for species diversification. This recent tendency to identify dispersals as drivers of biological diversification have increased with time (**Figure 1**). This tendency is related with the growing recognition of dispersal not only as a process able to generate congruent distributional patterns [11, 49, 61, 72], but also as a driver of species diversification [41]. It seems that the traditional and senseless dispute about whether vicariance or dispersal explains biotic distribution is a thing of the past [34, 49]. Both biogeographical events contribute to explain current and past geographic distributions and explain how biodiversity evolved in some regions. Accordingly, it is necessary to evaluate simultaneously which of these events were more prevalent through the biogeographic history of a taxa.

Theoretically, dispersal can promote exceptional shifts in species diversification by ecological opportunity (e.g., absence of related competitors) in a new colonized region or climate regime [1, 32, 33, 71, 73]. When lineages disperse to these new selective regimes they are likely to experience ecological release that can promote increases in speciation rates [33, 73]. Ecological opportunity can facilitate changes in diversification dynamics after a dispersal event [33, 73]. This is partially supported by evidence showing that dispersal events did not occur randomly across geography, and for some groups colonizing islands it was a notable increase in speciation rates after that events [4, 16, 21, 56, 57, 61]. For instance, dispersal of the palm tribe Trachycarpeae to different island systems promoted parallel diversifications connected with climatic and geological changes [4]. For bird clades distributed in the Indo-Pacific archipelago, species diversification rates increased on islands where few dispersal events were inferred, whereas clades occupying islands with high immigration rates exhibited constancy in diversification rates [16]. In the case

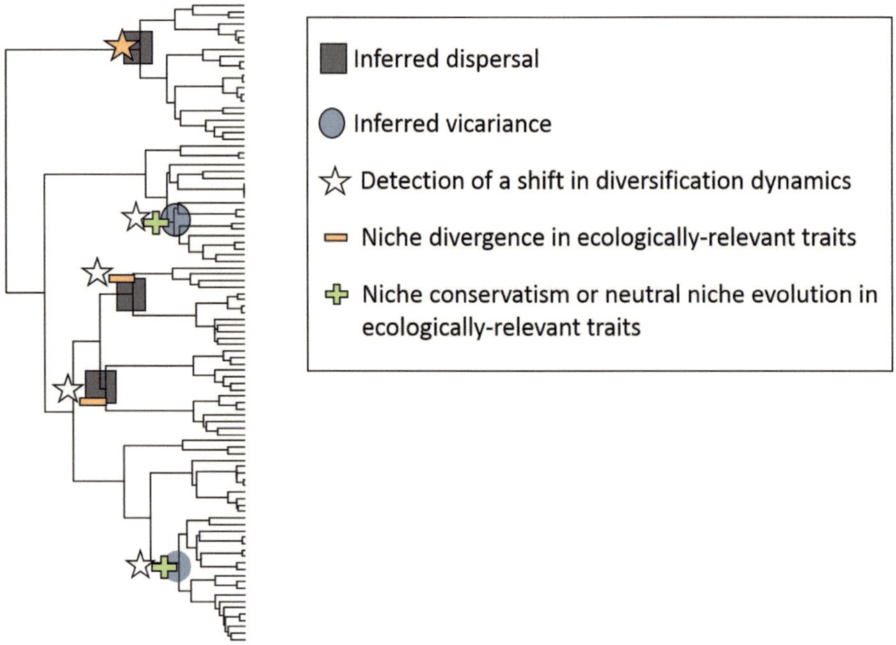

Figure 1. A hypothetical example showing inference of dispersal and vicariance and shifts in diversification dynamics across the phylogenetic diversification of a taxa. Nodes labeled with circles denote the position in the tree where it was inferred a dispersal (square) or a vicariance event (circles). Nodes labeled with stars denote where shifts in macroevolutionary regimes occurred. These shifts might involve increase in speciation rates or decrease in extinction rates or shifts in diversification dynamics from a time-dependent process to diversity-dependent cladogenesis (star). Under a dispersal scenario, where a lineage colonizes a new region with ecological opportunity, it is expected that ecologically relevant traits evolved through a pattern of niche divergence (dash). Whether all dispersal events promote speciation through ecological opportunity, it is expected that all events generated the same pattern of niche-trait divergence. By contrast, in a vicariance event, where the emergence of geographical barriers (e.g., a mountain uplift) promotes allopatric speciation, it is expected that ecologically relevant traits evolved little (i.e., phylogenetic niche conservatism) or under a neutral (drift) process in response to the changing environment (cross). Similarly, it is expected that all vicariant events generated the same pattern of niche conservatism or neutral evolution. Many other dispersal or vicariance events can be nested in the phylogeny; therefore, the inference of an event does not preclude that subsequent speciation events occur in sympatry or allopatry.

of Malagasy vangids, early dispersal from Africa to Madagascar promoted an increase in species diversification rates by ecological opportunity followed by a decline through time suggesting an ecological limit for this adaptive radiation [57].

Although these studies suggest a historical association between one or several dispersal events and shifts in net species diversification rates, it is still not clear which mechanisms were involved in this link between biogeography and macroevolution. Even more, it is not clear exactly how these past biogeographic events facilitate speciation in some regions and not in others or the mode of speciation involved in each event.

In contrast to the role of dispersal, vicariance has been associated as the only valid explanation for current distributional patterns and speciation for traditional biogeographers (e.g., Refs. [23, 45, 46]). However, few studies reviewed identified a vicariance event as a trigger of species diversification. Although these associations are congruent with the hypothesis of speciation by vicariance [38], it is still unknown whether the emergence of geographical barriers promoted habitat and reproductive isolation in taxa examined. For instance, rapid Andean uplift during the Miocene, with the concomitant climate changes, likely promoted allopatric speciation in *Heliotropium* plant lineages, which adapted to new arid conditions in southern South America [31]. Similarly, the Miocene uplift of the Qinghai-Tibetan plateau promoted allopatric speciation in Asteraceae clades plants [27] and *Rheum* plants [66]. These two clades of plants evolved as a response of temperature changes and the apparition of dry habitat conditions. It is likely that climate changes facilitated bursts of diversification [19, 27, 66]. Again, although these studies show an association between vicariance and species diversification, likely mediated through climate change or emergence of new barriers, it is not clear which mechanisms were involved here. All these studies suggest that we lack the ecological and evolutionary processes in these narrative biogeographic studies and we should look forward to identify at least which main processes underlying these historical associations.

It is evident that dispersal and vicariance events can be associated, either in a narrative or quantitative way, with shifts in species diversification. The link can be inferred directly in some cases [4, 41]; however, again the mechanisms operating behind these events are unknown. Although it is necessary to identify these evolutionary or ecological mechanisms, it is still not clear how these mechanisms can be disentangled. I consider that a full integration of current phylogenetic comparative methods with modern parametrical biogeographical methods and incorporation of ecologically relevant traits might offer new avenues to understand how biodiversity was generated after dispersal or vicariance events.

3.2. Limitations of current methods to link historical events with biological diversification

Some studies only provide a weak evidence of a consistent link between historical biogeographical events and shifts in species diversification rates. In particular, these few studies did not test whether dispersal or vicariance events effectively generated increases in net species diversification rates (e.g., Refs. [4, 7, 9, 68]. For instance, Chaves et al. [9] suggested that Andean uplift promoted increases in species diversification in the *Adelomyia* hummingbird genus. However, they only provide a correlation between divergence times for *Adelomyia* clades with some phases of the Andean uplift. Here, it is necessary to test whether Andean clades diversified more than other clades or the entire group. Accordingly, it is possible to suggest with higher confidence that dispersal to the Andes promoted posterior cladogenesis likely through allopatric speciation.

The link between historical biogeography and species diversification need to be explicitly tested using a two-step framework. First, it is necessary to infer historical biogeographic events with a higher confidence using powerful statistical methods [34]. These methods are ideal to identify rare dispersal events (founder-event speciation; [34]) and allow us to incorporate explicit geological information (e.g., timing of emergence of island landmasses; see Ref. [48] for an example using Caribbean *Anolis* lizards). Many empirical datasets show that founder-event speciation have left a strong imprint on distributional patterns of many taxa [34]. However, it is necessary to develop more complex models allowing the possibility to incorporate life-history traits and how these traits can affect rates of dispersal between regions (e.g., Ref. [35]). After the historical inference is conducted, it is crucial to identify whether these same nodes where a dispersal or vicariance event was inferred exhibit a notable shift in diversification dynamics [2, 3, 66].

Several methods allow to detect specific shifts in diversification dynamics in phylogenies, particularly increases in diversification rates [36, 37, 43, 53, 54]. These methods can be divided into temporal or topological methods, depending on the phylogenetic dataset at hand [43]. These methods also allow us to identify shifts in macroevolutionary regimes (i.e., shifts in diversification process; e.g., bayesian analysis of macroevolutionary mixtures (BAMM) approach developed by [53]; but see Ref. [42] for a criticism about the statistical power). However, these phylogenetic methods are not integrated fully with methods of historical biogeographic inference. Therefore, the association between dispersal or vicariance events and notables shifts in species diversification or diversification dynamics remains narrative [23]. Therefore, new methodologies are necessary to establish with more confidence whether these biogeographical events promote changes in biological diversification. Even more, it is possible that unmeasured traits (e.g., a "hidden" key innovation) affecting speciation and/or extinction rates evolve in the same nodes of a phylogenetic tree where a dispersal or vicariance event was inferred. In these cases, it might be very hard to distinguish whether diversification was affected by the presence (or absence) of a trait or by the movement to a new region (see Ref. [75]).

More robust methods to link historical biogeography and macroevolution are necessary. However, these new phylogenetic methods only can detect these historical associations, but these are very limited to establish the potential ecological or evolutionary process underlying this association. I consider that an integration of ecological studies (including testable predictions of trait-niche evolution) with a historical biogeographic approach will be useful to discern these mechanisms promoting speciation. The association between a biogeographic event and shifts in macroevolutionary dynamics is not enough to understand how biodiversity was generated in many regions. For instance, it has been found in warblers (*Phylloscopus*) that habitat isolation was the first form of niche differentiation to evolve in sympatry, followed by prey-size selection and feeding strategies [74]. It will be important to evaluate whether lineages diversifying after a dispersal event exhibit more niche differentiation [47] than lineages diversifying after a vicariant event. This can help to establish whether the initial trigger of speciation in a lineage was the colonization of a new region and the subsequent habitat isolation in sympatry [12, 13].

3.3. Is it possible to predict some ecological scenarios after a historical biogeographic event?

The ecological and evolutionary processes underlying the historical association between biogeography and diversification have been seldom discussed in the literature. Here, I propose that these contrasting historical events have left different imprints on the lineage diversification of taxa through its evolutionary history.

In a dispersal scenario, some lineages exposed to new ecological conditions (or ecological opportunity) likely experienced rapid diversification and rapid ecological divergence [18, 62, 64, 73]. In these cases, it is assumed that lineages evolved quickly by mechanisms of divergent selection due to ecological release and therefore occurs an increase in species diversification rates with ecological divergence. However, it is still not clear whether

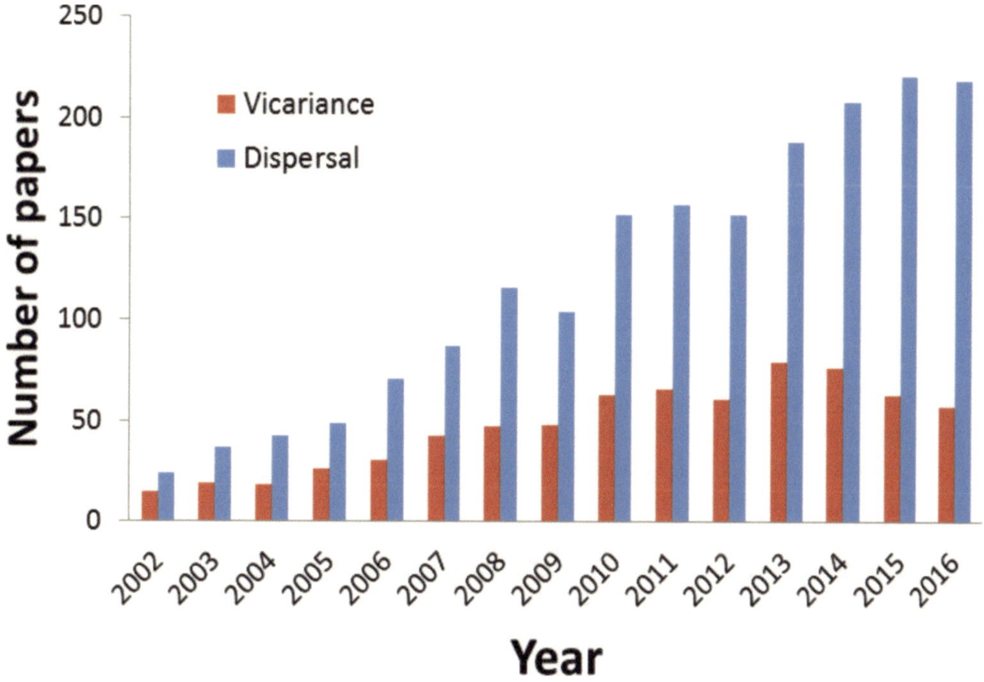

Figure 2. Number of papers published by year identifying dispersal or vicariance events as drivers of species diversification.

ecological or behavioral isolation occurred after the first lineage dispersed to a new region or speciation is driven by vicariance in the new colonized region (e.g., posterior mountain uplift in some islands or regions). Here, it is necessary to test whether a dispersal scenario promotes diversification in a similar fashion and therefore it is crucial to evaluate how species niche-traits evolved after a dispersal event. The colonization of a new region could not be the initial precursor of a change in macroevolutionary dynamics (e.g., increases in speciation rates) and other processes might be obscured here (e.g., reproductive isolation between ecologically diverging subpopulations inhabiting in sympatry [76]) (**Figure 2**).

By contrast, in a vicariance scenario I expected that all species diversification processes occur through allopatric divergence. In this scenario, new species emerge through geographical isolation of populations of a widespread ancestral species [12, 39]. Some studies reviewed here suggest that mountain uplift can generate a spatial gradient of climatic or ecological conditions that promote range subdivision and therefore allopatric speciation [5, 31]. However, still we do not know whether the mechanisms involved here will be different from a dispersal scenario. For instance, it is very well-known that ecological divergence between subpopulations is not a prerequisite for allopatric speciation. Therefore, species might acquire ecological niche differentiation by a pure genetic drift process [6] or species can exhibit strong niche stasis through time [26, 70]. Therefore, to establish whether vicariance effectively promote species diversification, it is crucial to test whether species evolved trait-niches through a pure drift process or maintain ancestral trait-niches due to stabilizing selection [6, 26, 70] (**Figure 2**).

In addition to this, climate change might act as a trigger of species diversification after a dispersal or vicariance event. For instance, many Neotropical diversification have been attributed to quaternary climatic cycles [77], where cooling phases in the Amazon basin facilitated geographic spread of highland species toward lowlands with subsequent range fragmentation and isolation during warm phases [77–79]. In the context of studies linking biogeographical events and diversification, very few studies have been able to detect a positive correlation between past climatic changes and species (or phenotypic) diversification after a given dispersal or vicariance event. For instance, Gamble et al. [80] suggested that main events of cladogenesis in *Gonatodes* geckos coincided with a phase of climate cooling in the late Eocence and Oligocene. In addition, Antonelli et al [81] suggested that speciation in an orchid South American group was accelerated by climate cooling after the Middle Miocene Climatic Optimum, 15 million years ago. However, these two studies did not evaluate whether specific shifts in net diversification rates matched temporally with significant decreases in temperature (i.e., cooling) in each region. Although these two studies suggest that climate change might be a trigger of species diversification, they did not mention how cladogenesis can be affected by climate changes, particularly cooling phases. The link between climate change and diversification is still narrative and more research is necessary to establish first whether there is possibility to find a positive association between both phenomena and then to establish the mechanisms driving speciation by climate change oscillations. For the first part, some recently developed models are able to detect whether paleo-environmental variations (e.g., temperature fluctuations through Cenozoic; [82]) had a significant effect on speciation and extinction rates [43, 83, 84]. These new models allow to establish whether the association is strong and how effectively climate can influence speciation and extinction rates for many groups [85].

Summarizing, both dispersal and vicariance scenarios can promote shifts in net species diversification rates in contrasting ways. In dispersal scenarios, it is expected as an early ecological niche differentiation through the lineage diversification of co-occurring species, as has been the case in Caribbean *Anolis* lizards [29] and Hawaiian spiders [17]. In vicariance scenarios, it is expected, that co-occurring species exhibit ecological niche conservatism in trait-niche axes, either in coarse-grain or fine-grain scales, or that these trait-niches likely evolved by a purely drift process. If ecological niche conservatism prevails after allopatric speciation, it is merely a by-product of the ancestral range fragmentation and therefore it is not possible that it promotes speciation (*contra* [25]). Otherwise, if there is some degree of niche differentiation between species this is likely due to either neutral evolution [6] or adaptation to a slowly changing environment [22]. The distinction between these two patterns (i.e., neutral versus adaptive niche evolution) will be crucial although it can be difficult to be detected using only phylogenetic comparative methods [44]. Furthermore, it is very likely that other ecological or non-ecological processes facilitating speciation occur in each one of these biogeographic scenarios [12]. This suggests that reproductive isolation between subpopulations as by-product of these processes (e.g., polyploidization, hybridization, and genetic drift) is not necessarily driven by historical dispersal or vicariance events. In addition, it is important to remember that inferring an early dispersal event (or vicariance) does not preclude that subsequent cladogenesis occurred by a successive series of vicariance events (or dispersals). For instance,

http://dx.doi.org/10.5772/intechopen.69516

a lineage colonized an oceanic island early in its evolutionary history, but the cladogenetic process that generated *in situ* diversity was driven by successive mountain uplifts.

3.4. Integration of a historical biogeography perspective with modern phylogenetic approaches and ecological studies

An integration of evidence from multiple study fields is necessary to get a deep understanding about the underlying mechanisms of the link between historical biogeography and diversification. Therefore, it will be crucial to develop and integrate new methodological approaches to generate sound evidence of a link between historical biogeographic events and changes in macroevolutionary dynamics. I consider that the first step necessary is to adopt a phylogenetic perspective to study the historical biogeography of a taxa and infer dispersal or vicariance events through its evolutionary history. Many studies reviewed have adopted such perspective, particularly implementing more complex event-based methods [60, 61]. New statistical biogeographical methods incorporating increasingly complex models of range evolution are being developed [34, 58]. The next step is to develop increasingly statistical robust methods to detect shifts in diversification dynamics and evaluate whether these shifts coincide with past dispersal or vicariance events and are not confounded by the emergence of hidden innovation key traits [53, 75]. Although there is a current debate about the power of these methods to detect these shifts [42, 53, 55], the BAMM approach ([53]) seems promising to detect these shifts in diversification dynamics across a phylogenetic tree with high confidence. BAMM allows us not only to detect shifts in speciation rates but shifts in the diversification dynamics itself. For instance, BAMM could allows us to detect in which nodes the diversification dynamics follows a time-dependent process, where only the time for speciation influences speciation rates [69]; or whether it follows a density-dependent cladogenetic process, where the clade diversity is limited by a carrying capacity [51]. Even more, using BAMM we could establish whether a macroevolutionary regime is more prone to occur.

In addition to the statistical detection of these historical association, I suggest that it is necessary to test the specific predictions that might emerge from each biogeographical scenario. I outlined two specific predictions for phenotypic trait evolutionary trajectories after a dispersal or vicariance event (see above). In that case, it is necessary to collect information about ecologically relevant traits for species to test these predictions. This information can be collected from museum specimens (e.g., morphological traits) or field-based studies (e.g., habitat use, diet or foraging strategy). Using model-based phylogenetic comparative methods, it is possible to evaluate whether these traits evolved following a Brownian motion or Ornstein-Uhlenbeck model [20] or whether traits evolved under a pure drift process lacking its phylogenetic signal (e.g., a white-noise process [20]). Furthermore, recently developed methods allow to identify shifts in phenotypic trait diversification in nodes of a phylogenetic tree [14, 15, 53, 59, 67]. In these cases, it is possible to evaluate simultaneously whether specific shifts in species or trait diversification evolved as a response to a dispersal or vicariance event. I consider that testing these predictions allows us to establish potential ecological and evolutionary mechanisms driving biological diversity. Some recent studies have begun to adopt a combination of these new methodologies (e.g., Ref. [24]), but still more research is necessary.

Finally, studies undoubtedly need to include detailed research from the field of the genetics of the adaptation to new environments [8, 22]. These studies are useful to understand why some clades diversified extensively after a dispersal or vicariance events and others not. For instance, species widely distributed in archipelagos or species with disjunct distributions will be good candidates for these detailed studies. Information about mechanisms maintaining or disrupting genetic cohesion between populations or lineages is crucial to understand the role of biogeographical events in the generation of biodiversity.

4. Conclusions

There is a growing interest in establishing the historical causes of biological diversification phenomena. However, we are still far behind to understand the process and mechanisms generating this realized diversification. In this chapter, I discussed some limitations of current methodologies employed to infer how biogeographical events as dispersal and vicariance can promote biological diversification shifts across the evolutionary history of lineages. The majority of studies reviewed here only were able to detect a historical association between a given biogeographic event and shifts in species (or phenotypic) diversification. Although we have established that spatial and temporal changes in diversification rates are modulated by abiotic (e.g., geography and climate) and biotic factors (e.g., food availability, predator presences, and intensity), still there is a lack of understanding about specific mechanisms underlying these historical associations and what factors were important. Furthermore, I have showed that the inference of these historical associations is not a trivial matter and more robust methodologies are necessary to establish a joint inference of biogeographical events and shifts in species diversification. In addition, I outlined specific predictions for evolutionary outcomes after a given biogeographical event which can be tested using current phylogenetic comparative methods and ecologically relevant traits. However, the adoption of phylogenetic methodological approaches will not be sufficient to generate a complete understanding of how species radiated after a given dispersal or vicariance event. Accordingly, traditional ecological and genetic field-based approaches used in speciation studies will be important here.

Acknowledgements

Funding was provided by a graduate scholarship grant from CONACYT and a postdoctoral research grant from DGAPA at Universidad Nacional Autónoma de México (UNAM). My ideas about biogeography and diversification were greatly benefited from discussions with Enrique Martinez Meyer, Oscar Flores Villela, Luna Sanchez Reyes and Mark. E. Olson. Thanks to Mark E. Olson to read several draft versions of this manuscript and for his continuous thoughtful comments. All remaining errors are my own.

Author details

Julián A. Velasco

Address all correspondence to: juvelas@gmail.com

Departamento de Biología Evolutiva, Facultad de Ciencias, Museo de Zoología 'Alfonso L. Herrera', Universidad Nacional Autónoma de México, Mexico city, Mexico

References

[1] Algar AC, Mahler DL. Area, climate heterogeneity, and the response of climate niches to ecological opportunity in island radiations of Anolis lizards. Global Ecology and Biogeography. 2015;**25**:781-791

[2] Almeida EAB, Pie MR, Brady SG, Danforth BN. Biogeography and diversification of colletid bees (Hymenoptera: Colletidae): Emerging patterns from the southern end of the world. Journal of Biogeography. 2011;**39**:526-544

[3] Antonelli A, Sanmartin I. Mass extinction, gradual cooling, or rapid radiation? Reconstructing the spatiotemporal evolution of the ancient angiosperm genus hedyosmum (Chloranthaceae) using empirical and simulated approaches. Systematic Biology. 2001;**60**:596-615

[4] Bacon CD, Baker WJ, Simmons MP. Miocene dispersal drives island radiations in the palm tribe Trachycarpeae (Arecaceae). Systematic Biology. 2012;**61**:426-42

[5] Blisniuk PM, Stern LA, Chamberlain CP, Idleman B, Zeitler PK. Climatic and ecologic changes during Miocene surface uplift in the Southern Patagonian Andes. Earth and Planetary Science Letters. 2005;**230**:125-142

[6] Boucher FC, Thuiller W, Davies TJ, Lavergne S. Neutral biogeography and the evolution of climatic niches. The American Naturalist. 2014;**183**:573-584

[7] Brumfield RT, Edwards SV. Evolution into and out of the Andes: A Bayesian analysis of historical diversification in Thamnophilus antshrikes. Evolution. 2007;**61**:346-367

[8] Campbell-Staton SC, Edwards SV, Losos JB. Climate-mediated adaptation after mainland colonization of an ancestrally subtropical island lizard, Anolis carolinensis. Journal of Evolutionary Biology. 2016;**29**:2168-2180

[9] Chaves JA, Weir JT, Smith TB. Diversification in Adelomyia hummingbirds follows Andean uplift. Molecular Ecology. 2011;**20**:4564-4576

[10] Cornell HV. Is regional species diversity bounded or unbounded? Biological Reviews of the Cambridge Philosophical Society. 2013;**88**:140-165

[11] Cowie RH, Holland BS. Dispersal is fundamental to biogeography and the evolution of biodiversity on oceanic islands. Journal of Biogeography. 2006;**33**:193-198

[12] Coyne JA, Orr HA. Speciation. Sinauer Associates, Sunderland, MA. 2004

[13] Dieckmann U, Doebeli M. On the origin of species by sympatric speciation. Nature. 1999;**400**:354-357

[14] Diniz-Filho JAF, Alves DMCC, Villalobos F, Sakamoto M, Brusatte SL, Bini LM. Phylogenetic eigenvectors and nonstationarity in the evolution of theropod dinosaur skulls. Journal of Evolutionary Biology. 2015;**28**:1410-1416

[15] Eastman JM, Alfaro ME, Joyce P, Hipp AL, Harmon LJ. A novel comparative method for identifying shifts in the rate of character evolution on trees. Evolution. 2011;**65**:3578-3589

[16] Fritz SA, Jønsson KA, Fjeldså J, Rahbek C. Diversification and biogeographic patterns in four island radiations of passerine birds. Evolution. 2012;**66**:179-190

[17] Gillespie R. Community assembly through adaptive radiation in Hawaiian spiders. Science. 2004;**303**:356-359

[18] Glor RE. Phylogenetic insights on adaptive radiation. Annual Review of Ecology, Evolution and Systematics. 2010;**41**:251-270

[19] Guo ZT, Ruddiman WF, Hao QZ, Wu HB, Qiao YS. Onset of Asian desertification by 22 Myr ago inferred from loess deposits in China. Nature. 2002;**416**:159-163

[20] Harmon LJ, Losos JB, Jonathan Davies T, Gillespie RG, Gittleman JL, Bryan JW, Purvis A. Early bursts of body size and shape evolution are rare in comparative data. Evolution. 2010;**64**:2385-2396

[21] Hedges B, Maxson R. Caribbean biogeography: Molecular evidence for dispersal in West Indian terrestrial vertebrates. Proceedings of the National Academy of Sciences. 1992;**89**:1909-1913

[22] Holt RD, Barfield M, Gomulkiewicz R. Temporal variation can facilitate niche evolution in harsh sink environments. The American Naturalist. 2004;**164**:187-200

[23] Humphries CJ, Parenti LR. Cladistic Biogeography: Interpreting Patterns of Plant and Animal Distributions. Oxford University Press, New York; 1999

[24] Kennedy JD, Borregaard MK, Jønsson KA, Holt B, Fjeldså J, Rahbek C. Does the colonization of new biogeographic regions influence the diversification and accumulation of clade richness among the Corvides (Aves: Passeriformes)? Evolution. 2016;**71**:38-50

[25] Kozak KH, Wiens JJ. Does niche conservatism promote speciation? A case study in North American salamanders. Evolution. 2006;**60**:2604-2621

[26] Kozak KH, Wiens JJ. Accelerated rates of climatic-niche evolution underlie rapid species diversification. Ecology letters. 2010;**13**:1378-1389

[27] Liu J-Q, Wang Y-J, Wang A-L, Hideaki O, Abbott RJ. Radiation and diversification within the Ligularia-Cremanthodium-Parasenecio complex (Asteraceae) triggered by uplift of the Qinghai-Tibetan Plateau. Molecular Phylogenetics and Evolution. 2006;**38**:31-49

[28] Losos JB. Adaptive radiation, ecological opportunity, and evolutionary determinism. The American Naturalist. 2010;**175**:623-639

[29] Losos JB, Jackman TR, Larson A, de Queiroz K, Rodríguez-Schettino L. Contingency and determinism in replicated adaptive radiations of island lizards. Science. 1998;**279**:2115-2118

[30] Losos JB, Schluter D. Analysis of an evolutionary species-area relationship. Nature. 2000;**408**:847-850

[31] Luebert F, Hilger HH, Weigend M. Diversification in the Andes: Age and origins of South American Heliotropium lineages (Heliotropiaceae, Boraginales). Molecular Phylogenetics and Evolution. 2011;**61**:90-102

[32] Mahler DL, Revell LJ, Glor RE, Losos JB. Ecological opportunity and the rate of morphological evolution in the diversification of Greater Antillean anoles. Evolution. 2010;**64**:2731-2745

[33] Mahler DL, Losos JB. Adaptive radiation: The interaction of ecological opportunity, adaptation, and speciation. In: Bell MA, Futuyma DJ, Eanes WF, Levinton JS, editors. Evolution Since Darwin: The First 150 Years. Sunderland, Massachusetts: Sinauer Associates; 2010. pp. 381-420

[34] Matzke NJ. Model selection in historical biogeography reveals that founder-event speciation is a crucial process in island clades. Systematic Biology. 2014;**63**:951-970

[35] Matzke NJ. Trait-dependent dispersal models for phylogenetic biogeography, in the R package BioGeoBEARS. Integrative and Comparative Biology. 2016;**56**:E330-E330

[36] May MR, Moore BR. How well can we detect Lineage-Specific Diversification-Rate shifts? A simulation study of sequential AIC methods. Systematic Biology. 2016;**65**:1076-1084

[37] May MR, Höhna S, Moore BR. A Bayesian approach for detecting the impact of mass-extinction events on molecular phylogenies when rates of lineage diversification may vary. Methods in Ecology and Evolution. 2016;**7**:947-959

[38] Mayr E. Systematics and the Origin of Species, from the Viewpoint of a Zoologist. Harvard University Press, Cambridge, MA; 1942

[39] Mayr E. Animal Species and Evolution. 1963. Cambridge University Press, Cambridge, MA; 1969

[40] Moore BR, Donoghue MJ. Correlates of diversification in the plant clade Dipsacales: Geographic movement and evolutionary innovations. The American Naturalist. 2007;**170**(Suppl):S28-55

[41] Moore BR, Donoghue. A Bayesian approach for evaluating the impact of historical events on rates of diversification. Proceedings of the National Academy of Sciences of the United States of America. 2009;**106**:4307-4312

[42] Moore BR, Höhna S, May MR, Rannala B, Huelsenbeck JP. Critically evaluating the theory and performance of Bayesian analysis of macroevolutionary mixtures. Proceedings of the National Academy of Sciences. 2016;**113**(34):9569-9574

[43] Morlon H. Phylogenetic approaches for studying diversification. Ecology Letters. 2014;**17**:508-525

[44] Münkemüller T, Boucher FC, Thuiller W, Lavergne S. Phylogenetic niche conservatism – common pitfalls and ways forward. Functional Ecology. 2014;**29**:627-639

[45] Nelson G, Platnick NI. Systematics and Biogeography: Cladistics and Vicariance. New York: Columbia University Press; 1981

[46] Parenti LR, Ebach MC. Comparative Biogeography. Discovering and Classifying Biogeographical Patterns of a Dynamic Earth. Berkeley and Los Angeles, California: University of California Press; 2009

[47] Peterson AT, Holt RD. Niche differentiation in Mexican birds: Using point occurrences to detect ecological innovation. Ecology Letters. 2003;**6**:774-782

[48] Poe S, Nieto-Montes de Oca A, Torres-Carvajal O, de Queiroz K, Velasco JA, Truett B, Gray LN, Ryan MJ, Köhler G, Ayala-Varela F, Latella I. A phylogenetic, biogeographic, and taxonomic study of all extant species of anolis (Squamata; Iguanidae). Systematic Biology. 2017. DOI: 10.1093/sysbio/syx029

[49] de Queiroz A. The resurrection of oceanic dispersal in historical biogeography. Trends in Ecology & Evolution. 2005;**20**:68-73

[50] Rabosky D, Glor R. Equilibrium speciation dynamics in a model adaptive radiation of island lizards. Proceedings of the National Academy of Sciences. 2010;**107**:22178-22183

[51] Rabosky DL. Ecological limits and diversification rate: Alternative paradigms to explain the variation in species richness among clades and regions. Ecology Letters. 2009;**12**:735-743

[52] Rabosky DL. Testing the time-for-speciation effect in the assembly of regional biotas. Methods in Ecology and Evolution. 2012;**3**:224-233

[53] Rabosky DL. Automatic detection of key innovations, rate shifts, and diversity-dependence on phylogenetic trees. PloS One. 2014;**9**:e89543

[54] Rabosky DL, Grundler M, Anderson C, Shi JJ, Brown JW, Huang H, Larson JG. BAMMtools: An R package for the analysis of evolutionary dynamics on phylogenetic trees. Methods in Ecology and Evolution. 2014;**5**:701-707

[55] Rabosky DL, Mitchell JS, Chang J. Is BAMM flawed? Theoretical and practical concerns in the analysis of multi-rate diversification models. Systematic Biology. 2017. DOI: 10.1093/sysbio/syx037

[56] Raxworthy CJ, Forstner MRJ, Nussbaum RA. Chameleon radiation by oceanic dispersal. Nature. 2002;**415**:784-787

[57] Reddy S, Driskell A, Rabosky DL, Hackett SJ, Schulenberg TS. Diversification and the adaptive radiation of the vangas of Madagascar. Proceedings of the Royal Society B. 2012;**279**:2062-2071. DOI: 10.1098/rspb.2011.2380

[58] Ree RH, Smith SA. Maximum likelihood inference of geographic range evolution by dispersal, local extinction, and cladogenesis. Systematic Biology. 2008;**57**:4-14

[59] Revell LJ, Mahler DL, Peres-Neto PR, Redelings BD. A new phylogenetic method for identifying exceptional phenotypic diversification. Evolution. 2012;**66**:135-146

[60] Sanmartin I. Event-Based biogeography: Integrating patterns, processes, and time. In: Ebach MC, Tangney R, editors. Biogeography in a Changing World. Taylor and Francis, Boca Raton, FL; 2007. pp. 135-156

[61] Sanmartín I, Van der Mark P, Ronquist F. Inferring dispersal: A Bayesian approach to phylogeny-based island biogeography, with special reference to the Canary Islands. Journal of Biogeography. 2008;**35**:428-449

[62] Simpson GG. Major Features of Evolution. New York: Columbia University Press; 1955

[63] Schemske D. Biotic interactions and speciation in the tropics. In: Butlin RK, Bridle J, Schluter D, editor. Speciation and Patterns of Diversity. Cambridge: Cambridge University Press; 2009. pp. 219-240

[64] Schluter D. The Ecology of Adaptive Radiaton. New York: Oxford University Press; 2000

[65] Schnitzler J, Graham CH, Dormann CF, Schiffers K, Peter Linder H. Climatic niche evolution and species diversification in the Cape flora, South Africa. Journal of Biogeography. 2012;**39**:2201-2211

[66] Sun Y, Wang A, Wan D, Wang Q, Liu J. Rapid radiation of Rheum (Polygonaceae) and parallel evolution of morphological traits. Molecular Phylogenetics and Evolution. 2012;**63**:150-158

[67] Thomas GH, Freckleton RP. MOTMOT: Models of trait macroevolution on trees. Methods in Ecology and Evolution. 2012;**3**:145-151

[68] Upham NS, Patterson BD. Diversification and biogeography of the Neotropical caviomorph lineage Octodontoidea (Rodentia: Hystricognathi). Molecular Phylogenetics and Evolution. 2012;**63**:417-429

[69] Wiens JJ. The causes of species richness patterns across space, time, and clades and the role of "ecological limits". The Quarterly Review of Biology. 2011;**86**:75-96

[70] Wiens JJ, Graham CH. Niche conservatism: Integrating evolution, ecology, and conservation biology. Annual Review of Ecology and Systematics. 2005;**36**:519-539

[71] Wüest RO, Antonelli A, Zimmermann NE, Linder HP. Available climate regimes drive niche diversification during range expansion. The American Naturalist. 2015;**185**:640-652

[72] Yoder AD, Nowak MD. Has vicariance or dispersal been the predominant biogeographic force in madagascar? Only time will tell. Annual Review of Ecology, Evolution, and Systematics. 2006;**37**:405-431

[73] Yoder JB, Clancey E, Des Roches S, Eastman JM, Gentry L, Godsoe W, Hagey TJ, Jochimsen D, Oswald BP, Robertson J, Sarver BAJ, Schenk JJ, Spear SF, Harmon LJ. Ecological opportunity and the origin of adaptive radiations. Journal of Evolutionary Niology. 2010;**23**:1581-1596

[74] Richman AD, Price T. Evolution of ecological differences in the Old World leaf warblers. Nature. 1992;**355**:817

[75] Beaulieu JM, O'Meara BC. Detecting hidden diversification shifts in models of trait-dependent speciation and extinction. Systematic Biology. 2016;**65**:583-601

[76] Nosil P. Ecological Speciation. Oxford University Press, New York; 2012

[77] Rull V. Neotropical biodiversity: Timing and potential drivers. Trends in Ecology & Evolution. 2011;**26**:508-513

[78] Noonan BP, Gaucher P. Phylogeography and demography of Guianan harlequin toads (Atelopus): Diversification within a refuge. Molecular Ecology. 2005;**14**:3017-3031

[79] Noonan BP, Wray KP. Neotropical diversification : The effects of a complex history on diversity within the poison frog genus Dendrobates. Journal of Biogeography. 2006;**33**:1007-1020

[80] Gamble T, Simons AM, Colli GR, Vitt LJ. Tertiary climate change and the diversification of the Amazonian gecko Gonatodes (Sphaerodactylidae, Squamata). Molecular Phylogenetics and Evolution. 2008b;**46**:269-277

[81] Antonelli A, Verola CF, Parisod C, Gustafsson ALS. Climate cooling promoted the expansion and radiation of a threatened group of South American orchids (Epidendroideae: Laeliinae). Biological Journal of the Linnean Society. 2010;**100**:597-607

[82] Zachos J, Pagani M, Sloan L, Thomas E, Billups K. Trends, rhythms, and aberrations in global climate 65 Ma to present. Science. 2001;**292**(5517):686-693

[83] Condamine FL, Rolland J, Morlon H. Macroevolutionary perspectives to environmental change. Ecology Letters. 2013;**16**(s1):72-85

[84] Morlon H, Lewitus E, Condamine FL, Manceau M, Clavel J, Drury J. RPANDA: An R package for macroevolutionary analyses on phylogenetic trees. Methods in Ecology and Evolution. 2016;**7**:589-597

[85] Hua X, Wiens JJ. How does climate influence speciation? The American Naturalist. 2013;**182**(1):1-12

Chapter 3

Ecological Responses to Climate Change at Biogeographical Boundaries

Melinda Pálinkás

Additional information is available at the end of the chapter

http://dx.doi.org/10.5772/intechopen.69514

Abstract

Temporal and spatial biogeographical boundaries are usually associated with extinction processes. However, some boundary regions seem to be places of speciation. It is unclear if boundaries are favored by generalized or specialized species. Recent studies suggest that narrow-ranging species can be strong competitors and they can replace wide-ranging species while shifting their range boundaries under the effect of climate change. In other boundary regions, the decline of both passive- and active-dispersing specialists has been observed. Core regions are also weakening. They are jeopardized mainly by extreme climate events and fragmentation and by the accompanied non-native invasions. Biodiversity loss and homogenization have been observed globally.

Keywords: biogeographical boundary, ecotone, core region, climate change, mass extinction, specialist, generalist

1. Introduction

Temporal and spatial biogeographical patterns change in space and time. Historical biogeographical boundaries usually mark great mass extinction events. The role of spatial boundaries is controversial in recent global changes. Some suggest that species at sharp biogeographical boundaries are at the edge of their existence and doomed to extinction, others allege that transition zones can serve as biodiversity hot spots. They harbor either wide-ranging species adapted to broad environmental circumstances or narrow-ranging species occurring in unique environments. Researchers assuming specialized species in boundary regions imply that biogeographical boundaries are suitable for climate change detection and specialists can be used as early warning signals. Several studies suggest that specialization is the greatest extinction risk [1]. However, mass extinction events affect both specialized and generalized species. Studies show that both groups are declining under global changes. Interactions between specialists and generalists are not exactly clear. It is urgent to detect their locations globally and clarify their roles. It is also an alarming trend that not only boundaries but also core regions are weakening, which leads to homogenization, the abundance of generalized species, and biodiversity loss. Recent anthropogenic changes are complex including not only climatic changes but also habitat destruction, fragmentation, and pollution which act synergistically.

This chapter addresses the following issues: (1) Are biogeographical boundaries the scenes of extinction? (2) Which factors weaken spatial boundaries and core regions? (3) Are core areas threatened by climate change? (4) Are biogeographical boundaries unique regions? (5) Do they harbor generalized or specialized species? (6) What are the roles of specialists and generalists in extinction processes?

2. Temporal biogeographical pattern

Temporal and biogeographical boundaries cannot be separated from each other. Temporal boundaries eliminate old spatial boundaries and create new ones. Temporal boundaries are usually associated with extinction processes. It is suggested that mass extinctions start at local scales and spill over to higher scales in time [2].

2.1. Succession

Local species and communities replace each other in time. At ecological time scale, this process is induced by repeated disturbances, and it is called succession. The Clementsian school considers succession as a deterministic process that culminates in a predictable stable, "climax" community [3]. Succession is not random, because it is determined by climatic and soil conditions. In contrast, the stochastic Gleasonian school suggests that a single region can have several successional stages at the same time and more than one stable stages or climaxes [1].

2.2. Mass extinction

Major extinction events indicate boundaries in geological time. Approaching an extinction event, ecosystems display specific traits serving as warning signals of a catastrophic shift. (Post-extinction periods also show distinctive pattern of biotic restructuring.) Extinction events as temporal boundaries eliminate old spatial boundaries and create new ones. Non-catastrophic extinctions affect biological systems at different spatial scales and different trophic levels in a selective way. The inherent extinction proneness of taxa also contributes to the selective nature of extinctions. In contrast, catastrophic extinction events or, in other words, mass extinctions affect the whole global ecosystem in a non-selective way wiping out most living creatures. Several studies suggest that we are undergoing the sixth mass extinction.

Non-selectivity is the main characteristic that makes a difference between background and mass extinction. During mass extinction events, widespread and abundant species also extinct

[4, 5]. The disappearance of generalists is a sign of shifting toward nonselectivity [6, 7]. (Large body as a main extinction trait is often mentioned in literature; therefore, losing large-body mammals [8] is an early indicator.)

At global spatial scale and at longer time period (historical time scale), sudden and large environmental perturbations wipe out whole biotas causing mass extinction. This large-scale, repeated replacement is similar to local succession. Apart from mass extinctions, changes in biotas are of smaller magnitude and rather gradual. That is why an increase in frequency and magnitude of changes in communities or biotas is an early signal of a regime shift. At geological time scale, mass extinctions usually mark a boundary between time units (e.g., eras, period, epochs), the tipping point of a biotic shift. They are associated with drastic environmental perturbations (sudden climate change, volcanism, sea-level changes, meteor impact events). Referring back to recent climatic changes, historical mass extinctions accompanied with global warming can provide valuable information for us to be able to presage future trends.

3. Spatial biogeographical pattern

The geographic ranges of species evolve under limited environmental conditions creating a spatial pattern. Broad-ranging species perceive fewer boundaries than species with restricted geographic ranges, and they can shift their ranges relatively more easily under changing environmental conditions.

Spatial boundaries are affected by natural biotic and abiotic factors and anthropogenic disturbation which enhance each other's effect through interactions. Extreme changes in these factors and in the inherent traits of boundaries can lead to extinctions.

3.1. Abiotic factors

The abundance and the distribution of species are usually affected by the synergy of multiple environmental factors, such as temperature, water availability, soil and water chemistry, etc. For example, the tolerance of high temperature is typically lower in plants, which don't tolerate decreased soil moisture. Local extinctions at the boundaries of species ranges are common during droughts [1, 9, 10].

3.1.1. Extreme perturbation

Disturbances such as fires, storms, and volcanic eruptions either destroy or maintain boundaries, depending on their magnitude and frequency. Natural ignition (lightning), for example, prevents woody encroachment and exotic species invasion at forest and shrub/grassland boundaries; therefore, artificial fire suppression leads to forest expansion. In arid regions, the decrease in natural *fires* coupled with livestock grazing often results in desertification. Desert shrublands expand at the expense of grasslands [1, 11, 12]. Synergistic processes have an important role in this case as well. Fragmentation lowers the probability of lightning-ignited fires. Increased fragmentation along with the disappearance of an important boundary regulator leads to the local extinctions of native grassland species which can spill over to higher spatial levels supporting the homogenization processes.

The investigations conducted by du Toit et al. [13] in the South African Nama Karoo transition zone confirmed that more frequent and/or more intensive fires can lead to a biome shift if the most abundant species fails to recover after an extreme disturbance in a transition zone. The dominant vegetation of Nama Karoo is grass, and shrub and fires are rare. They monitored the recovery of the vegetation after a natural ignition. Most of the species managed to recover except the most abundant Karoo shrub species seven months after the fire. This might suggest a biome shift from shrubland to grassland.

3.1.2. *Extreme weather pattern*

Climate change enhances the magnitude and the frequency of extreme events [14]. Frequent extreme climatic events, e.g., extreme droughts, weaken both core areas and boundary regions by altering species composition, diversity, and functional and structural attributes. Native species being less adaptive to extreme events may be displaced by non-native generalist invaders [15].

Boundaries are more exposed to extreme events than core regions; therefore, even the events of low magnitude can degrade their structure. Several studies confirm that relatively weak winds can contribute to the invasion of weedy species by dropping wind-transported seeds at the edges [16, 17].

Recent *droughts* have induced forest canopy thinning in the core areas of tropical forests. In some high-rainfall places, forests have disappeared probably because of the relatively long dry season in Australia [18, 19]. Longer dry periods have also been experienced in tropical montane forests in Costa Rica with severe consequences [20]. Drier climatic conditions opened a path for pathogenic invaders from lower altitudes [21] resulted in the die-off of most endemic frog and toad species during the 1980s [22]. This example illustrates the devastating effects of synergistic extinction drivers on endemic species. According to Fjeldså [23], the lack of endemic species in a tropical montane forest indicates that the local biotic community cannot maintain a hydrological balance anymore and withstand global changes.

3.1.3. *Habitat destruction and fragmentation*

Habitat destruction and fragmentation can be considered as extreme anthropogenic perturbation. Fragmentation is detrimental for specialized species. It eliminates intact core zones and reduces the imperviousness of edges providing open space for non-native, wide-ranging species. The higher trophic level and large body size make terrestrial species sensitive to fragmentation. This can further enhance the extinction proneness of African megaherbivores maintaining biome boundaries.

Janzen [24] confirmed that fragmentation leads to *weed expansion* in habitat patches. Forest fragmentation results in smaller patches which probably become more and more distinct from the intact forest, because the mortality of native tree species along the edges is higher than that of environmentally more tolerant weedy species [24]. The success of weed invasion depends on the width and the imperviousness of buffer zones as well as their relative dispersal abilities [25]. Buffer zones are the zones between the core areas and edges, or, in another point of view, they can be considered as wider edge zones. If they are occupied by weedy species, native interior tree species cannot reestablish [26]. In small patches, forest specialists can be completely replaced by generalists after perturbation [27]. Conservationists emphasize that it is important to preserve larger habitat patches which presumably contain more specialist species. Nevertheless, Beier et al. [28] pointed out that the generalists inhabiting small habitat patches provide important ecosystem services; therefore, they can be the centers for future ecosystem recovery [28].

In general, higher trophic levels give stronger responses to fragmentation and habitat loss than lower trophic levels [29–31]. Krauss et al. [32] assume that lower population sizes, higher population variability, and dependence on lower trophic levels are the main reasons for fragmentation susceptibility of higher trophic levels. Large body size can also enhance the sensitivity to fragmentation and increase the extinction risk of terrestrial species according to several sources [33].

3.2. Biotic factors

The main biotic factors forming boundaries are *competition, predation, and mutualism*.

3.2.1. Competition

Species limit each other's distribution by *competition*. Strong competition can result in non-overlapping range boundaries [1]. Non-overlapping boundaries display sudden regime shifts under environmental changes. The current shifting of species ranges is also influenced by competition, which affects both the generalized and specialized species.

In the last decades, woody encroachment has been experienced globally under the effects of global warming [34–36] mainly because of CO_2 enrichment. Woody species which are generally superior competitors [1, 37] tend to be sensitive to abiotic stress (fire, drought). However, they experienced fewer detrimental perturbations recently, which also helped their expansion.

The relationship between species diversity and geographic range limitation affects spatial patterns [1]. Abiotic and biotic factors vary along range boundaries. Under unfavorable environmental conditions, species diversity and hence competition are lower. When environmental conditions are beneficial for most species, diversity increases and biotic interactions (e.g., competition, predation) will become the limiting factors. This might be the reason why many biodiversity hot spots are located along the tropical biome boundaries.

The global spatial pattern of generalist and specialist species reflects the changing abiotic conditions in a similar way. In the tropical zone where the environmental conditions are favorable, the diversity and the biotic interactions are high, many species tend to be specialized, and the ecosystems are productive. Proceeding to the poles, environmental conditions become more unfavorable, diversity and productivity decrease, and the species become more generalized. Isolated and small geographic ranges (small islands and forest fragments) are also homogenized and dominated by a few generalized species because of the unfavorable conditions. Decreasing geographic ranges and increasing disturbance jeopardize both specialized and generalized species.

The tropical region provides interesting examples for diffuse competition which also modifies species ranges under recent climate change. Proceeding to the equator, the southern limits of the geographical ranges become less climate dependent and more effective by competition in the Northern Hemisphere. MacArthur et al. [38] suggest that strong biotic competition restricts some tropical species to habitats with less favorable environmental conditions. The same species can turn into widespread and abundant species in subtropical and temperate zones by diffuse competition. Yellow warbler (*Dendroica petechia*) is a good example for that. Its geographic range is widely expanded in the temperate zone, while under tropic conditions, it is strongly restricted [38]. According to MacArthur et al. [38], diffuse competition of tropical species is on increase.

Bennett et al. [37] also observed strong tropical competitors in the temperate zone. Tropical herbivorous fish shifted northward at the expense of seaweeds. Seaweeds are dominant, wide-spreading taxa in subtropical and temperate coastal zones. The poleward shift of tropical herbivorous fish prevents the recovery of seaweeds and maintains a canopy-free alternative state after the extreme disturbances (overgrazing).

In some cold regions, specialists are displacing generalized species. Directional taxonomic shifts of the algal communities in the Northern Hemisphere have been observed by Ruhland et al. [39], especially in the alpine regions and arctic zones with a tendency of an increase in specialized taxa which are replacing generalized species [40].

3.2.2. Predation

Predation can limit the distribution of both predators and preys. Specialization or overhunting can lead to a drop in prey abundance, and this way both groups suffer. The geographical ranges of highly specialized predators are usually further constricted by other limiting factors; hence, they are especially prone to extinction.

3.2.3. Mutualism

Mutualism results in the identical ranges of parasites and hosts; therefore, coevolved species at boundaries and in core regions are prone to co-extinction. Mutualism-related co-extinction is strongly enhanced by fragmentation. Co-extinction affects both specialist and generalists, which can lead to wider extinction.

Grasslands are endangered globally. Grassland specialists can expect a long-term decline because of the drastic loss of their habitats [32]. Time-delayed extinction of long-lived vascular plants may bring about the co-extinction of short-lived specialized herbivores, e.g., butterflies [32].

Rainforests are also jeopardized by habitat destruction. The decline of old native trees in rainforests because of fragmentation may cause the co-extinction of specialized mutualists and herbivores [41].

Invasion can replace core super-generalists in the mutual networks, as well. Giannini et al. [42] observed invasive super-generalist bee species in Brazil replacing native super-generalist species which can modify the interactions in networks. The non-native, super-generalist bee species invaded into the core of the networks rapidly. Romanuk et al. [43] and Lurgi et al. [44] suggest that large and more generalist species are the best invaders.

Dario Palacio et al. [45] studied a highly diverse network of plant and fruit-eating birds in a cloud forest in the Colombian Andes. They found that the elimination of super-generalists which are the connectors of disconnected subsets of species makes the mutualistic network prone to collapse despite its high diversity. They experienced the early decline of large frugivores forming the core of the network because of their high vulnerability to fragmentation. They also noted that the early loss of endemic and specialized species may precede the decline of central super-generalists. However, the extinction of less-connected specialized species presumably does not lead to the collapse of the whole network in contrast with the decline of the central super-generalist species. Similar networks are located in the Atlantic Forest in Brazil as well which are also threatened by extinction [46]. The authors' results suggest that generalist species play an important role in the ecosystem functions.

3.2.4. Dispersal abilities

Both active- and passive-dispersing specialist species are declining. Specialist species are at great risk even if they are active dispersal.

Good dispersals are able to shift their ranges and avoid abiotic stress. For this reason, the natural range boundaries of plants and sessile animals change relatively slowly. For instance, the contemporary biome distribution pattern in Africa does not reflect the actual current climate but historical conditions [47].

According to Terborgh [48], mainly specialization, high trophic level, and poor dispersal ability promote extinction. Laurance [49] and Turner et al. [50] suggest that mammals and plants with poor dispersing abilities are more prone to extinction than active dispersers, which leads to a higher abundance of generalist species [51]. Wilson and Willis [51] highlight the early loss of specialists during extinction events. Short-lived pollinators with good dispersal abilities shifted their ranges in North America and Europe under climate change [52]. Short-lived specialists are sensitive to environmental changes [53], which makes them good early warning indicators of perturbation. Bartomeus et al. [54] described a decline in plant-pollinator networks throughout the US over the last 120 years. Scheffers et al. [55] suggest that specialized pollination systems are expected to be more vulnerable and hence more sensitive indicators of global warming. Krauss et al. [32] found that short-lived specialist butterflies experienced severe decline after perturbation despite the fact that they are active dispersers.

Rare species are usually more localized, sparse, and relatively more specialized [33]. Their geographic ranges are more fragmented; hence, metapopulation and edge effects can be significant contributors of their decline [56]. Specialized taxa tend to be rare, which increases the extinction likelihood [33]. Rarity and specialization are two different traits, but they often act synergistically. However, Didham et al. [57] pointed out that range-restricted species may be more disperse and persistent than common, sessile species in small fragments. Didham et al. [57] investigated the effects of forest fragmentation on beetle species in central Amazonia. They found that rare species were better survivor in small fragments than "common" species. They concluded that rare species are more mobile and more persistent in contrast with competitively dominant but more sessile species which are more prone to extinction under forest fragmentation. Hanski and Ovaskainen [58] argue that the transient abundance of rare species can be experienced after excessive habitat loss and fragmentation.

3.2.5. Sensitive development stages of species

Species at different development stages show different tolerance of environmental conditions, which affects their range sizes, their boundary types, and boundary perception. For example, the life cycle of a frog or a dragonfly includes very different ranges and boundaries because of the varied niches of stages.

Higgins et al. [59] emphasize that the growth rate of Savanna tree seedling and saplings affects their survival during fire events. Fire suppression, especially during the sensitive development stages of trees, favors woody encroachment.

3.2.6. Continental drift

The theory of continental drift was formed during the last century. It was a revolutionary step, and it revealed the secret of several vague biogeographical issues, for example, the omnipresence of sessile animals, which are not able to cross oceans. Plate tectonics is responsible for the birth and the destruction of continents. The assemblage and the positions of continents are changing. Their union creates bridges between terrestrial biotas providing free gene flow, and their separation may lead to their isolation. These processes are selective as species are sensitive to boundaries to different degrees especially considering their dispersal abilities, but it can be stated that global changes of large magnitude affect most species uniformly in many cases.

According to Lyell's geoclimatic theory, the concentration of continents near the equator triggers global warming, while the juxtaposition of landmasses close to the poles evokes global cooling. Hence, continental drift can be considered as a climate regulator and thus a temporal boundary "creator."

The collision of continental plates can establish a connection between biotas, but paradoxically it can create a spatial boundary as well, since continental collisions produce towering mountain ranges which are restrictive to lowland species. The union of landmasses is a violent event erasing and reshaping boundaries. The Great Permian Extinction may have also been associated with the formation of the Pangea supercontinent which brought about a significant drop in the sea level and the drying of the continental shelves [1]. However, Pangea also served as a cradle for many survivors and novel species which expanded their range boundaries over the continent. When the continents separated, global climatic conditions changed dramatically again. The species survived this event radiated and diversified under new environmental circumstances.

4. Some important spatial traits of biogeographical boundaries

Biogeographical boundaries can be categorized in many ways [60, 61]. Here, mainly sharp boundaries are discussed in relation to global changes.

Controversial views on boundaries are partly generated by incoherent spatial scales applied in studies.

4.1. Spatial scale

Climate has a great effect on the biogeographical pattern. Geography and meteorology apply similar spatial scales which makes the scientific investigations more consistent. Saunders and Briggs [62] emphasize the importance of proper scale. If biogeographical problems are not managed at the proper scale, it can lead to the loss of biota. The mismatches of human-related and natural boundaries can deteriorate the environment. Improper scale also brings about biased and controversial data.

Sub-local spatial scale (< a few meters) includes microhabitats and small boundaries. For example, the boundaries between surfaces of different exposures on a boulder also mark the borders between the patches of different lichens. Local spatial scale (a few meters to 1 km) deals with the level of communities. Regional spatial scale (1–100 km) can be related to landscape boundaries, and continental spatial scale (>100 km) is appropriate for researches on landmass boundaries. Increasing spatial scale is usually associated with increasing temporal scale, from a couple of hours or days to millions of years.

4.2. Spatial origin: natural vs. anthropogenic

Natural boundaries are the formations of the nature which divide two or more different units of natural origin, like timberlines, mountain chains, and watercourses. Anthropogenic boundaries are usually man-made objects (transportation, industrial, residential elements) and the boundaries of anthropogenic plant communities (croplands and plantations). Anthropogenic boundaries are always sharp representing an obstacle or filter to migration and gene flow. They can be either physical objects or boundaries of high contrast between the adjacent units, for instance, edges between forests and croplands where different microclimatic and ecological conditions meet.

4.3. Spatial structure: sharp vs. gradual

Nature can produce relatively quick changes at boundaries as well; however, along environmental gradients, abiotic and biotic changes are gradual. This leads to an important difference between sharp and gradual biogeographical boundaries. They are usually referred to as "ecotones" and "ecoclines" in ecology.

Starting with the latter one, ecoclines are ecosystems in which the associated communities show a gradual change along an environmental gradient. The environmental heterogeneity results in gradual phenotypic and/or genetic differences of species which are also called ecotypes. This gradual variation reflects an adaption to the changing environment. In an ecocline the physiological characteristics of plants and animals change gradually proceeding to higher latitudes (e.g., the skin color in human populations). This phenomenon can lead to speciation only if the environmental conditions change dramatically.

Researchers usually show more interest in ecotones which represent sharp biogeographical boundaries between ecosystems.

5. Ecotones

Sharp boundaries are usually referred to as ecotones in literature. It is suggested that sharp boundaries (hereinafter ecotones) might be unique environments.

Ecotones have been studied for more than a century [63–65]; however, researchers have devoted more attention to the investigation of distinct, relatively homogeneous ecological units until recently. Various authors suggest that understanding boundaries may have an important role in the early detection of global climate change [66–70] and in conservation works [71–74].

Ecotones are also referred to as transition zones, junction zones, tension belts, edges, borders, etc. Ecotones can be considered as the edge or the periphery of an ecological system or as a transient zone between two or more adjoining ecological units. Ecological boundaries which have sharp environmental and ecological gradients are usually unstable [75]. They share common traits with the adjoining regions but also hold unique features [76]. Ecotones promote high biodiversity and unique, rare, specialized, vulnerable species, which make them biodiversity hot spots [74] and may be central regions for future conservation efforts.

Ecotones harbor range-restricted species which are mostly considered to be vulnerable to climatic changes and fragmentation and thus prone to extinction. According to researchers, specialists will be the first to extinct under the sixth mass extinction. The role of specialists prior to extinction processes has a main priority in most studies as they can be used as early warning signals. Generalists as the main survivors of environmental changes are usually disregarded in approaching havocs, though they maintain the communities as well. Kark and van Rensburg [74] argue that not only ecotones but also core regions are threatened by global changes.

Kark and van Rensburg [74] raised an important research question related to ecotonal species assemblage: "Are they young species currently diverging in the ecotone region via parapatric speciation or rather wide-ranging species that have expanded their ranges to ecotonal environments?" Studies are controversial in this respect, and they emphasize the importance of both generalist and specialist species in core regions and in boundary regions as well.

5.1. Generalized and specialized species in ecotones

In literature, wide-ranging species are implied to in many ways, such as generalist, generalized, widespread, abundant, r-strategist, weed, ruderal, tolerant, invasive, opportunistic, pioneer, and widely dispersing. Narrow-ranging species are referred to as range restricted, narrowly adapted, specialist, k-strategist, competitive, endemic, rare, unique, vulnerable, sensitive, etc. Generalized species are able to adapt to a broad variety of environmental conditions, and they can shift their diet. Specialists are less flexible in adaptation, and they occupy only a narrow range of niche.

Gosz [77] suggests that edge species are likely to be generalist, wide-ranging, and dominant. Generalists are able to cross boundaries. Wide-ranging, generalist taxa are more mobile than sensitive, vulnerable taxa which tend to be sessile; that's why generalists perceive fewer boundaries and detect the landscape more homogenous [78, 79]. Generalist can be forced to leave their habitat and cross boundaries by habitat destruction or overpopulation. For instance, wide-ranging predators leave overpopulated habitat patches and cross the boundaries in cross-edge spillover predation [80–82].

Some studies suggest that generalists might have an important role both in core regions and at boundaries by maintaining communities. For example, krill have an important role in connecting different trophic levels in oceans. They are widespread globally; however, Antarctic krill occur only along the boundary between sea ice and ocean water, because they can find both rich food and shelter from predators there [83].

According to traditional textbooks, specialized species tend to become rare or even lost in a deteriorated environment. In contrast, generalist species prefer impaired habitats where they are found in great number. Disturbed and damaged sites are occupied by generalist species adopting disturbance strategy. However, ecotones can be under disturbance, still having lots of specialized species, and damaged tropical grasslands are rich in specialists as well.

Others studies suggest that the unique environmental conditions favor specialized and endemic species in ecotones [71, 84].

According to Morelli [85], both specialists and generalists should be applied as bioindicators in disturbed landscapes because of the homogenization of communities. He used bird observation data to identify avian hot spots. He selected specialized species in natural environments and both generalized and specialized species in disturbed environments. The selected species varied in different environments. He found that only a few common species are enough to detect high species richness hot spots. He also observed that two specialized bioindicators occurred both in cultivated and natural landscapes (in forest and in grassland, respectively).

McKinney [33] points out that extinction promoting traits tend to covary. According to Brown's hypothesis [86], species having narrow niche are adapted narrowly in several parameters, whereas species with broader niche are broadly adapted in not only one but several parameters. Furthermore, narrow niche is characterized by low local abundance and small geographical range [87, 88]. Considering the synergistic combination of traits related to narrow niche, the fate of specialist species is sealed under anthropogenic threats [87, 88].

Generalists are usually broadly adapted in not only one but several parameters, while specialists are narrowly adapted in many respects [87, 88] so they represent two extremes of adaptation and thus two extremes of extinction proneness. However, it is important to note that the degree of specialization and generalization can urge or delay extinction processes in the transition zones and in the core regions as well.

Broadly adapted biotas are able to shift their ranges in response to climatic changes [89]. Biotas which are broadly adapted can keep pace with global warming more easily and may experience lower rate of extinction. Several paleontological records confirm the extinction resistance traits of generalist species [90–92]. Generalist species are more resistant to background and mass extinction than specialist ones. Mammals are more specialized than insects, and small mammals are more generalized than large mammals [93]. Scheffers et al. [55] evaluated literature on climate change impacts. They concluded that warming climate may result in a decreased body size in most cases as a large surface-to-volume ratio is more favorable under warm climate [94].

Despite the long history of ecotone investigations [63, 65], studies show mixed results on the role of transition zones in maintaining high diversity [95]. Odum [76] suggested among the first ones that ecotones may have high species richness and unique, endemic species. Since then, several studies seem to have confirmed that near ecotones, species richness and rarity are increased. Kark and van Rensburg [74] claim that boundary regions sustain high diversity because of the adjoinings and overlapping ecoregions (mass effect), but they are also locations for speciation and hence rare and unique species. Kark et al. [95] found that passerine birds, including rare species, occur in higher number in transition zones than in the adjacent ecoregions in the New World. van Rensburg et al. [96] concluded that range-restricted birds and frogs are frequently located closer to ecotones in South Africa. Kark [95] pointed out that rainforest ecotones in Central Africa may be the centers of speciation as a result of evolutionary and ecological processes, hence supporting the biodiversity of the whole biome. Kark et al. [97, 98] observed a biodiversity hot spot at a sharp ecotone between the Mediterranean and semi-arid regions in southern Israel. It is important to note that rarity is one of the best predictors of extinction [33, 91, 99].

Biogeographic regions with the significant level of biodiversity and high rate of endangered species are considered as biodiversity hot spots. It is an interesting question if biodiversity hot spots are ecotonal or rather core regions. The tropical zone is the most abundant of biodiversity hot spots. It has approximately ten times more biodiversity hot spots than the non-tropical zones do [100]. Stevens [101] claims that tropical species are generally more endemic and smaller and they have narrower ranges than temperate species, which make them extinction prone. This might suggest that in the tropical zone both core areas and ecotones have an important role in maintaining biodiversity. Several studies suggest that future extinction will affect the humid tropics the most severely [102, 103].

Tropical grasslands are also diverse and rich in endemic species, and they are as endangered as forests. Grassy biomes include biodiversity hot spots with lots of endemic species. Non-forest habitats are rich in endemic vertebrates and invertebrates. Non-forests hold 30–50% of plant diversity [104]. Ancient grasslands which are alternative stable states of forests are probably rich in endemic species. For example, Cerrado tropical grassy biome in Brazil is a threatened biodiversity hot spot [105].

High rainfall grasslands in Brazil [106], Africa [107], Thailand [108], etc. have a particularly high level of plant diversity and many endemic species. The Indian montane grasslands have many endemic species [109]. Madagascan grasslands are also rich in endemics [110, 111].

Grassy biomes have high light requirements and disturbance tolerance. The similar may be true for sharp boundaries between tropical grasslands and forests. These boundaries are maintained by megaherbivores and fires. High diversity and high number of specialized (and endemic) species are typical for grasslands. Open savannas labeled as "disturbed" or "degraded" harbor many specialists and maintain high diversity in Madagascar and Indonesia [104]. Grassland fauna resists to fire and has great resilience. Savanna species are usually competitive, are mobile, and have a wide range of diet, which means that they can shift their diet, and they prefer open environments [112]. Bond and Parr [104] allege that the loss of grassland specialist birds can be used as early warning signals of shifts to forest at landscape scale considering their large habitat requirements. According to Skowno and Bond [113], specialized bird species of different levels of forest already appeared in significant number in grassy ecosystems.

According to Strayer et al. [114], species assemblage and interactions along boundaries may be unique, or they may represent the average of the adjacent patches. They refer to these two types as "interactive and noninteractive boundaries." Under certain circumstances, ecotones may be unique environments separately from the adjoining communities and not the mix of the adjacent environments.

5.2. Ecotones and climate change

The Earth's climate can be characterized by natural cycles of cooling and warming phases. Cooling usually results in less diverse and broadly adapted biotas with selectively eliminated tropical biotas. Warming is beneficial for the development of more complex and specialized biotas [91]. Currently, we are in a controversial situation. Despite the fact that we are undergoing a natural cooling process lowering the diversity level, we are experiencing anthropogenic global warming, which also contributes to extinctions because of its high rate.

The role of ecotones in climate change processes is unclear. Gaston et al. [115] suggest that ecotones are sensitive to global warming as ecotonal species are already at the edge of their ranges, which make them prone to extinctions. Others argue that ecotones are places of temporal and spatial fluctuations; hence, ecotonal communities should be more resistant to global warming [74]. Some also suggest that changes in ecotones might serve as early warning signals of ecosystem shifts [50, 51]. Ecotones may be viable areas that sustain themselves over time, or they are temporary product of constant flow from the adjacent communities [116]. This might have an effect on their persistence to future global changes.

5.2.1. Importance of ecotones in mass extinction

Conservation works have shifted from protecting of individuals to identifying regions with high diversity [117]: botanical hot spots [118] and hot spots of endemic birds [119], which are targets of mass extinction as rare species are concentrated in small areas. We can assume that a part of the biodiversity hot spots might be transition zones, some of which are rich in young and novel species. Brooks and McLennan [120] and Erwin [121] propose that these regions will be the first victims of mass extinction as they contain restricted-range species in small place so they can be wiped out completely. On the other hand, they might be also the centers of repopulation after mass extinction.

5.2.2. Low latitude ecotones as future refugia

Hampe and Petit [122] suggest that southern (rear) edge of species ranges should deserve greater attention or at least should not be neglected compared to the more studied northern (poleward) expanding edge, as the rear-edge populations store the species' genetic diversity. This might be applied as analogue in case of greater transition zones serving as biodiversity hot spots. It is an interesting question whether low latitude transitional zones are the most important biodiversity hot spots serving as a refugium in future mass extinction.

Based on the estimation of the Late Quaternary glacial-interglacial climate displacement rate, Sandel et al. [123] concluded that high-velocity and unstable regions tend to have mainly widespread species which are resilient to climatic oscillations and have strong dispersal abilities. Their results show that during the Late Quaternary the northeastern part of North America and the north-central Eurasia had the highest velocity and the weakly dispersing amphibians were affected the most. They pointed out that low-velocity regions can be refuges for sessile and small-ranged species [123]. Many bird and mammal endemic species are concentrated in the Southern Hemisphere where a higher velocity of changes can be expected according to predictions [123].

6. Discussion

Biogeographical boundaries are shifting globally. Late Quaternary glacial-interglacial climate change proves that climate displacement rate tends to vary regionally [123]. Sandel et al. [123] argues that high-velocity and unstable regions have mainly widespread species which are resilient to climatic oscillations and have strong dispersal abilities. However, the rapid expansion of specialized species has been observed in the tropical, temperate, and arctic zone as well as in the mountains [1, 55]. Warming climate seems to favor species with strong competitive and dispersal abilities. Recent studies [55] suggest that non-sessile specialized species

http://dx.doi.org/10.5772/intechopen.69514

which are strong competitors thrive in high-velocity, shifting boundary regions and as Brown and Lomolino [1] conclude that they start to behave as generalists. Other studies describe the extinction of both active- and passive-dispersing specialized species [32]. Short-lived pollinators and birds, for instance, are at great risk.

According to Sandel et al. [123], low latitude transitional zones harbor sessile, small-ranged species and can be characterized by low climate displacement rate. He suggests that low-velocity regions might serve as refuges under anthropogenic extinction processes. Sandel et al. [123] predict that the climate displacement rate will be higher in the Southern Hemisphere than it was during the Late Quaternary climate change. The Southern Hemisphere is rich in endemic hot spots, which suggests a higher rate of endangerment and biodiversity loss. It can also mean that regions which could serve as refugia might be exterminated. Tropical grassland and forest biomes and their boundary regions maintain high diversity and rich in endemic species; therefore, they are jeopardized by global warming.

Several studies pointed out that some ecotones are biodiversity hot spots and they are places for speciation. These observations originate mainly from the tropical and subtropical zones [72, 95]. The core regions harbor specialized species as well. This raises important questions. What are the roles of core region and boundary specialists in extinctions and how much they differ (if they differ) in extinction proneness? Many studies claim that specialization is one of the greatest extinction risks [33], which makes specialized species good bioindicators. Can core region specialists expand their ranges under global warming or they are among the first victims because of the weakening core regions? As nothing is black and white, maybe no obvious answer exists. Local and regional divergences as well as the synergy of many factors suggest several outcomes. For example, African megaherbivores are considered to be specialized in diet. However, recent studies [124] show that they can shift their diet, which makes them more generalized than previously thought. Still, they are endangered boundary species mainly because of overhunting and habitat destruction. Their large body size and higher tropic level also contribute to extinction proneness.

Some studies [77] claim that generalized species might be the beneficiaries of climate change as they are more adaptive than specialized species. However, specialists are displacing generalized species which are supposed to be weaker competitors in many places. Native super-generalists are being expelled by invasive super-generalists in mutualist networks. The decay of generalized species is a threatening issue, because they maintain communities. Fragmentation is a key contributor of their decline in many cases. The increasing number of perishing specialized and generalized species probably refers to a post-initial phase of mass extinction. Morelli [85] suggests the use of both specialists and generalists as bioindicators in deteriorated regions.

Zhu et al. [34] and others observed woody encroachment in many regions all over the world, which might suggest that it helps maintain biodiversity. However, it jeopardizes grassland biodiversity hot spots. Even degraded tropical grasslands harbor several rare, endemic, specialized species. Fragmentation and fewer numbers of natural fires also contribute to the decay of grasslands. At the same time, tropical forests, paradoxically, are also suffering. Extreme perturbations affect not only boundary but also core regions, which can trigger the invasion of exotic species and the extinction of native species. Climate change–induced woody encroachment is not necessarily accompanied by an increase in biodiversity. On the contrary, biodiversity loss is detected worldwide.

In summary, climate change affects most levels of the global ecosystem. Both core regions and boundaries are eroding which leads to biodiversity loss and homogenization. Decaying generalized species probably refer to a post-initial stage of mass extinction.

Author details

Melinda Pálinkás

Address all correspondence to: m.plinka@gmail.com

Szent István University, Budapest, Hungary

References

[1] Brown JH, Lomolino MV. Biogeography. 2nd ed. Sunderland, Massachusetts: Sinauer Association, Inc. Publishers; 1998. 691 p

[2] Merriam G, Wegner J. Local extinctions, habitat fragmentation, and ecotones. Hansen A, di Castri F, editors. In: Landscape Boundaries: Consequences for Biotic Diversity and Ecological Flows. New York: Springer-Verlag; 1992. pp. 150-159

[3] Odum EP. The strategy of ecosystem development. Science. 1969;**164**(3877):262-270

[4] Jablonski D. Lessons from the past: Evolutionary impacts of mass extinctions. Proceedings of the National Academy of Sciences of the United States of America. 2001;**98**(10):5393-5398

[5] Jablonski D. Colloquium paper: Extinction and the spatial dynamics of biodiversity. Proceedings of the National Academy of Sciences of the United States of America [Internet]. 2008;**105**(Supplement_1):11528-11535. Available from: http://www.pnas.org/content/105/Supplement_1/11528.abstract

[6] Ceballos G, Ehrlich PR. Mammal population losses and the extinction crisis. Science [Internet]. 2002;**296**(5569):904-907. Available from: http://www.sciencemag.org/cgi/doi/10.1126/science.1069349

[7] Cardillo M. Multiple causes of high extinction risk in large mammal species. Science. 2005;**309**(5738):1239-1241

[8] Isaac NJB, Turvey ST, Collen B, Waterman C, Baillie JEM. Mammals on the EDGE: Conservation priorities based on threat and phylogeny. PLoS One. 2007;**2**(3):1-7

[9] Sinclair W. Comparison of recent declines of white ash, oaks and sugar maple in northeastern woodlands. Corneell Plant. 1964;**20**:62-67

[10] Westing AH. Sugar maple decline. An evaluation. Economic Botany. 1966;**20**:196-212

[11] Johnston MC. Past and present grasslands of southern Texas and northeastern Mexico. Ecology. 1963;**44**:456-466

[12] Bahre CJ. Human impacts on the grasslands of Southeastern Arizona. McClaran M, Van Devender TR, editors. In: The Desert Grassland. Tucson: University of Arizona Press; 1995. pp. 230-264

[13] du Toit JC, van den Berg L, O'Connor TG. Fire effects on vegetation in a grassy dwarf shrubland at a site in the eastern Karoo, South Africa. African Journal of Range and Forage Science [Internet]. 2014;**119**(March 2015):1-8. Available from: http://www.tandfonline.com/doi/abs/10.2989/10220119.2014.913077

[14] Edenhofer O, Pichs-Madruga R, Sokona Y, Minx JC, Farahani E, Kadner S, et al, editors. IPCC Report [Internet]. Climate Change 2014: Mitigation of Climate Change. Part of the Working Group III Contribution to the Fifth Assessment Report of the Intergovernmental Panel on Climate Change. IPCC; 2014. 1-161 p. Available from: http://ebooks.cambridge.org/ref/id/CBO9781107415416A011

[15] Jiménez M, Jaksic F, Armesto J, Gaxiola A. Extreme climatic events change the dynamics and invasibility of semi-arid annual plant communities. Ecology [Internet]. 2011 [cited 2016 Nov 6]; Available from: http://onlinelibrary.wiley.com/doi/10.1111/j.1461-0248.2011.01693.x/full

[16] Duncan DH, Dorrough J, White M, Moxham C. Blowing in the wind? Nutrient enrichment of remnant woodlands in an agricultural landscape. Landscape Ecology. 2008;**23**(1):107-119

[17] Hirota M, Holmgren M, Van Nes EH, Scheffer M. Global resilience of tropical forest and savanna to critical transitions. Science [Internet]. 2011;**334**(6053):232-235. Available from: http://www.ncbi.nlm.nih.gov/pubmed/21998390

[18] Hirota M, Holmgren M, Nes EV, Scheffer M. Global resilience of tropical forest and savanna to critical transitions. Science [Internet]. 2011 [cited 2016 Nov 6]; Available from: http://science.sciencemag.org/content/334/6053/232.short

[19] Liedloff AC, Cook GD. Modelling the effects of rainfall variability and fire on tree populations in an Australian tropical savanna with the FLAMES simulation model. Ecological Modelling. 2007;**201**(3):269-282

[20] Pounds AJ, Fogdon MPL, Cambell JH. Biological response to climate change on a tropical mountain. Nature. 1999;**398**:611-615

[21] Seimon TA, Seimon A, Daszak P, Halloy SRP, Schloegel LM, Aguilar CA, et al. Upward range extension of Andean anurans and chytridiomycosis to extreme elevations in response to tropical deglaciation. Global Change Biology. 2007;**13**(1):288-299

[22] Pounds JA, Bustamante MR, Coloma LA, Consuegra JA, Fogden MPL, Foster PN, et al. Widespread amphibian extinctions from epidemic disease driven by global warming. Nature [Internet]. 2006;**439**(7073):161-167. Available from: http://www.ncbi.nlm.nih.gov/pubmed/16407945

[23] Fjeldså J. Geographical patterns for relict and young species of birds in Africa and South America and implications for conservation priorities. Biodiversity and Conservation [Internet]. 1994;**3**:207-226. Available from: http://link.springer.com/article/10.1007/BF00055939%5Cnhttp://link.springer.com/content/pdf/10.1007/BF00055939.pdf

[24] Janzen DH. The eternal external threat. Soule ME, editor. In: Conservation Biology: The Science of Scarcity and Diversity. Massachusetts: Sinauer Associates, Inc.; 1986. pp. 286-303

[25] Davies-Colley R, Payne G, Elswijk MV. Microclimate gradients across a forest edge. New Zealand Journal [Internet]. 2000 [cited 2016 Nov 7]; Available from: http://www.jstor.org/stable/24054666

[26] Janzen DH. No park is an island: Increase in interference from outside as park size decreases. Oikos. 1983;**41**:402-410

[27] Ambuel B, Temple SA. Area dependent changes in the bird communities and vegetation of southern Wisconsin USA forests. Ecology. 1983;**64**:1057-1068

[28] Beier P, Van Drielen M, Kankam BO. Avifaunal collapse in West African forest fragments. Conservation Biology. 2002 Aug;**16**(4):1097-1111

[29] Didham RK, Hammond PM, Lawton JH, Eggleton P, Stork NE. Beetle species responses to tropical forest fragmentation. Ecological Monographs. 1998;**68**:295-323

[30] Gilbert F, Gonzalez A, Evans-Freke I. Corridors maintain species richness in the fragmented landscapes of a microecosystem. Proceeding of the Royal Society of London B: Biological Science [Internet]. 1998;**265**(1396):577-582. Available from: http://rspb.royalsocietypublishing.org/cgi/doi/10.1098/rspb.1998.0333

[31] Komonen A, Penttilä R, Lindgren M, Hanski I. Forest fragmentation truncates a food chain based on an old-growth forest bracket fungus. Oikos. 2000;**90**:119-126

[32] Krauss J, Bommarco R, Guardiola M, Heikkinen RK, Helm A, Kuussaari M, et al. Habitat fragmentation causes immediate and time-delayed biodiversity loss at different trophic levels. Ecology Letters. 2010;**13**(5):597-605

[33] McKinney ML. Extinction vulnerability and selectivity: Combining ecological and paleontological views. Annual Review of Ecological System [Internet]. 1997;**28**(1):495-516. Available from: http://www.annualreviews.org/doi/10.1146/annurev.ecolsys.28.1.495

[34] Zhu Z, Piao S, Myneni RB, Huang M, Zeng Z, Canadell JG, et al. Greening of the Earth and its drivers. Nature Climate Change. 2016;**6**(August):1-6 early online

[35] Garamvölgyi Á, Hufnagel L. Impacts of climate change on vegetation distribution no. 1: Climate change induced vegetation shifts in the palearctic region. Applied Ecology and Environmental Research. 2013;**11**(1):79-122

[36] Hufnagel L, Garamvoelgyi A. Impacts of climate change on vegetation distribution no. 2—Climate change induced vegetation shifts in the new world. Applied Ecology and Environmental Research. 2014;**12**(2):355-422

[37] Bennett S, Wernberg T, Harvey ES, Santana-Garcon J, Saunders BJ. Tropical herbivores provide resilience to a climate-mediated phase shift on temperate reefs. Ecology Letters. 2015;**18**:714-723

[38] MacArthur RH, Diamond JM, Karr JR. Density compensation in island faunas. Ecology. 1972;**53**:330-342

[39] Ruhland K, Patterson AM, Smol JP. Hemispheric-scale patterns of climate-related shifts in planktonic diatoms from North American and European lakes. Global Change Biology. 2008;**14**:1-15

[40] Thienpont JR, Korosi JB, Cheng ES, Deasley K, Pisaric MFJ, Smol JP. Recent climate warming favours more specialized Cladoceran taxa in western Canadian Arctic lakes. Journal of Biogeography. 2015;**42**(8):1553-1565

[41] Laurance SG, Gomez MS. Clearing width and movements of understory rainforest birds. Biotropica. 2005;**37**:149-152

[42] Giannini TC, Garibaldi LA, Acosta AL, Silva JS, Maia KP, Saraiva AM, et al. Native and non-native supergeneralist bee species have different effects on plant-bee networks. PLoS One [Internet]. 2015; Available from: http://dx.doi.org/10.1371/journal.pone.0137198

[43] Romanuk TN, Zhou Y, Brose U, Berlow EL, Williams RJ, Martinez ND. Predicting invasion success in complex ecological networks. Philosophical Transactions of the Royal Society B Biological Sciences [Internet]. 2009;**364**:1743-1754. Available from: http://www.pubmedcentral.nih.gov/articlerender.fcgi?artid=2685429&tool=pmcentrez&rendertype=abstract

[44] Lurgi M, Lopez B, Montoya J. Novel communities from climate change. Philosophical Transactions of the Royal Society B Biological Sciences. 2014;**367**:2913-2922

[45] Dario Palacio R, Valderrama-Ardila C, Kattan GH. Generalist species have a central role in a highly diverse plant-frugivore network. Biotropica. 2016 May;**48**(3):349-355

[46] Vidal MM, Hausi E, Pizo MA, Tamashiro JY, Silva WR, Guimaraes PR. Frugivores at higher risk of extinction are the key elements of a mutualistic network. Ecology. 2014;**95**:3440-3447

[47] Moncrieff GR, Bond WJ, Higgins SI. Revising the biome concept for understanding and predicting global change impacts. Journal of Biogeography. 2016;**43**(5):863-873

[48] Terborgh J. Preservation of natural diversity: The problem of extinction prone species. Bioscience. 1974;**24**:715-722

[49] Laurance WF. Rainforest fragmentation and the structure of small mammal communities in tropical Queensland. Biological Conservation. 1994;**69**(1):23-32

[50] Turner JRG, Lennon JJ, Greenwood JJD. Does climate cause the global biodiversity gradient? Hochberg M, Claubert J, Barbault R, editors. In: Aspects of the Genesis and Maintenance of Biological Diversity. Oxford, UK: Oxford University Press; 1996. p. 1996

[51] Wilson EO, Willis EO. Applied biogeography. In: Cody ML, Diamond JM, editors. Ecology and Evolution of Communities. Cambridge, MA: Belknap Press; 1975. pp. 522-534

[52] Burkle LA, Marlin JC, Knight TM. Plant-pollinator interactions over 120 years: Loss of species, co-occurrence and function. Science. 2013;**339**(6127):191-193

[53] Morris WF, Pfister CA, Tuljapurkar S, Haridas CV, Boggs CL, Boyce MS. Longevity can buffer plant and animal populations against changing climatic variability. Ecology. 2008;**89**:19-25

[54] Bartomeus I, Ascher JS, Gibbs J, Danforth BN, Wagner DL, Hedtke SM, et al. Historical changes in northeastern US bee pollinators related to shared ecological traits. Proceedings of the National Academy of Sciences [Internet]. 2013;**110**(12):4656-4660. Available from: http://www.ncbi.nlm.nih.gov/pubmed/23487768

[55] Scheffers BR, DeMeester L, Bridge TCL, Hoffmann AA, Pandolfi JM, Corlett R, et al. The broad footprint of climate change from genes to biomes to people. Science. 2016;**354**(6313):719

[56] Maurer B, Nott M. Geographic range fragmentation and the evolution of biological diversity. McKinney M, editor. In: Biodiversity Dynamics: Turnover of Populations, Taxa, and Communities. New York: Columbia University Press; 1997. pp. 31-50

[57] Didham R, Hammond P, Lawton J. Beetle species responses to tropical forest fragmentation. Ecological Monographs 68: 295-323 [Internet]. 1998 [cited 2016 Nov 4]; Available from: http://onlinelibrary.wiley.com/doi/10.1890/0012-9615(1998)068[0295:BSRTTF]2.0.CO;2/full

[58] Hanski I, Ovaskainen O. Extinction debt at extinction threshold. Conservation Biology. 2002;**16**(3):666-673

[59] Higgins SI, Bond WJ, Trollope WSW. Fire, resprouting and variability: A recipe for grass—tree coexistence in savanna. Journal of Ecology. 2000;**88**:213-229

[60] Banks-Leite C, Ewers RM. Ecosystem boundaries. In: Encyclopedia of Life Sciences. Chichester, UK: John Wiley & Sons, Ltd; 2009. pp. 1-8

[61] Strayer D, Power M, Fagan W. A Classification of Ecological Boundaries. BioScience, 53(8):723-729. 2003 [cited 2016 Nov 4]; Available from: http://bioscience.oxfordjournals.org/content/53/8/723.short

[62] Saunders DA, Briggs SV. Nature grows in straight lines—Or does she? What are the consequences of the mismatch between human-imposed linear boundaries and ecosystem boundaries? An Australian example. Landscape and Urban Planning. 2002;**61**:71-82

[63] Clements FE. Plant succession: An analysis of the development of vegetation. Washington: Carnegie Institution of Washington; 1916. pp. 140-143

[64] Livingston BE. The distribution of the upland plant societies of Kent County, Michigan. Botanical Gazette. 1903;**35**(1):36-55

[65] Weaver JE, Thiel AF. Ecological Studies in the Tension Zone between Prairie and Woodland; The Botanical Survey of Nebraska. New Series. Number I. The University of Nebraska, Lincoln. 1917. 60 p

[66] Solomon AM. Transient response of forests to CO_2-induced climate change: Simulation modeling experiments in eastern north America. Oecologia [Internet]. 1986;**68**(4):567-579. Available from: http://www.jstor.org.proxy-remote.galib.uga.edu/stable/4217884

[67] Sipkay C, Horváth L, Nosek J, Oertel N, Vadadi-Fülöp C, Farkas E, et al. Analysis of climate change scenarios based on modelling of the seasonal dynamics of a Danubian copepod species. Applied Ecology and Environmental Research. 2008;**6**(4):101-109

[68] Diós N, Szenteleki K, Ferenczy A, Gergely P, Levente H. A climate profile indicator based comparative analysis of climate change scenarios with regard to maize cultures. Applied Ecology and Environmental Research. 2009;**7**(3):199-214

[69] Hansen A, Risser P, di Castri F, Hansen AJ, Risser PG, di Castri F. Epilogue: Biodiversity and ecological flows across ecotones. Hansen A, di Castri F, editors. In: Landscape Boundaries: Consequences for Biotic Diversity and Ecological Flows. New York: Springer-Verlag; 1992. pp. 423-438

[70] Allen CD, Breshears DD. Drought-induced shift of a forest-woodland ecotone: Rapid landscape response to climate variation. Proceedings of National Academy Sciences [Internet]. 1998;**95**(25):14839-14842. Available from: http://www.pnas.org/content/95/25/14839.short

[71] Smith TB, Wayne RK, Girman DJ, Bruford MW. A role for ecotones in generating rainforest biodiversity. Science. 1997;**276**:1855-1857

[72] Smith TB, Kark S, Schneider CJ, Wayne RK, Moritz C. Biodiversity hotspots and beyond: The need for preserving environmental transitions. Trends in Ecology & Evolution. 2001;**16**(8):431

[73] Brooks T, Hannah L, Da Fonseca GAB, Mittermeier RA. Prioritizing hotspots, representing transitions [2]. Trends in Ecology & Evolution. 2001;**16**(12):673

[74] Kark S, van Rensburg BJ. Ecotones: Marginal or central areas of transition? Israel Journal of Ecology and Evolution [Internet]. 2006;**52**(1):29-53. Available from: http://www.tandfonline.com/doi/abs/10.1560/IJEE.52.1.29

[75] Kent M, Gill W, Weaver R. Landscape and plant community boundaries in biogeography. Progress in Physical Geography. 1997; 21(3): 315-353 [cited 2016 Nov 7]; Available from: http://ppg.sagepub.com/content/21/3/315.short

[76] Odum EP. Fundamentals of Ecology. Philadelphia: W.B. Saunders Company; 1953. 383 p

[77] Gosz JR. Ecological functions in a biome transition zone: Translating local responses to broad-scale dynamics. Hansen A, di Castri F, editors. In: Landscape Boundaries: Consequences for Biotic Diversity and Ecological Flows. New York: Springer-Verlag; 1992. pp. 55-75. Available from: http://dx.doi.org/10.1007/978-1-4612-2804-2_3

[78] Wiens J, Crawford C, Gosz J. Boundary dynamics: A conceptual framework for studying landscape ecosystems. Oikos [Internet]. 1985 [cited 2016 Nov 4];**45**(3):421-427. Available from: http://www.jstor.org/stable/3565577

[79] Fahrig L. Effect of spatial arrangement of habitat patches on local population size. Fahrig L, Paloheimo J, editors. In: Ecological Society of America Stable. Effect of Spatial. 2012;**69**(2):468-475. http://www.jstor.org/stable/1940445

[80] Holt RD. Population dynamics in two-patch environments: Some anomalous consequences of an optimal habitat distribution. Theoretical Population Biology. 1985;**28**(2): 181-208

[81] Oksanen T. Exploitation ecosystems in heterogeneous habitat complexes. Evolutionary Ecology. 1990;**4**(3):220-234

[82] Cantrell RS, Cosner C, Fagan WF. How predator incursions affect critical patch size: The role of the functional response. The American Naturalist [Internet]. 2001;**158**(4):368-375. Available from: http://www.ncbi.nlm.nih.gov/pubmed/18707333

[83] Brierley AS, Fernandes PG, Brandon MA, Armstrong F, Millard NW, Mcphail SD, et al. Antarctic Krill under sea ice: Elevated abundance in a narrow band just south of ice edge. Source Science News Series [Internet]. 2002;**295**(5561):1890-1892. Available from: http://www.jstor.org/stable/3076223%0Ahttp://about.jstor.org/terms

[84] Ferro I, Morrone JJ. Biogeographical transition zones: A search for conceptual synthesis. Biological Journal of the Linnean Society. 2014;**113**(1):1-12

[85] Morelli F. Indicator species for avian biodiversity hotspots: Combination of specialists and generalists is necessary in less natural environments. Journal for Nature Conservation. 2015;**27**:54-62

[86] Brown JH. Macroecology. Chicago: Chicago University Press; 1995. 284 p

[87] Lawton JH. Population dynamic principles. Lawton JH, May RM, editors. In: Extinction Rates. Oxford, UK: Oxford University Press; 1995. pp. 147-163

[88] Lawton J, Nee S, Letcher A, Harvey P. Animal distributions: Patterns and processes. In: Edwards P, May R, Webb N, editors. Large-Scale Ecology and Conservation Biology. New York: Blackwell; 1994. pp. 41-58

[89] Coope G. Insect faunas in ice age environments: Why so little extinction?. In: Lawton J, May R, editors. Extinction Rates. Oxford: Oxford University Press; 1995. pp. 55-74

[90] McGhee G. The Late Devonian Mass Extinction. New York: Columbia Univ. Press; 1996. 302 p

[91] Stanley S. Delayed recovery and the spacing of major extinctions. Paleobiology 1990;**16**:401-414

[92] Van Valen LM. Concepts and the nature of natural selection by extinction: Is generalization possible?. In: Glen W, editor. The Mass Extinction Debates. Palo Alto, USA: Stanford Univ. Press; 1994. pp. 200-216

[93] Calder W. Size, Function and Life History. Cambridge, MA: Harvard Univ. Press; 1984. 387 p

[94] Sheridan JA, Bickford D. Shrinking body size as an ecological response to climate change. Nature Climate Change [Internet]. 2011;**1**(8):401-6. Available from: http://www.nature.com/doifinder/10.1038/nclimate1259

[95] Kark et al. 2007. The role of transitional areas as avian biodiversity centres. Global Ecology and Biogeography. 2007;**16**(2):187-196

[96] van Rensburg BJ, Levin N, Kark S. Spatial congruence between ecotones and range-restricted species: Implications for conservation biogeography at the sub-continental scale. Diversity and Distributions. 2009;**15**(3):379-389

[97] Kark S, Alkon PU, Safriel UN, Randi E. Conservation priorities for chukar partridge in Israel based on genetic diversity across an ecological gradient. Conservation Biology. 1999;**13**(3):542-552

[98] Kark S, Mukerji T, Safriel UN, Noy-Meir I, Nissani R, Darvasi A. Peak morphological diversity in an ecotone unveiled in the chukar partridge by a novel Estimator in a Dependent Sample (EDS). The Journal of Animal Ecology. 2002;**71**(6):1015-1029

[99] Mace G, Kershaw M. Extinction risk and rarity on an ecological timescale. Kunin W, Gaston K, editors. In: The Biology of Rarity. London, UK: Chapman & Hall; 1997. pp. 130-149

[100] Brook BW, Sodhi NS, Bradshaw CJA. Synergies among extinction drivers under global change. Trends in Ecology & Evolution. 2008;**23**(8):453-460

[101] Stevens GC. The latitudinal gradients in geographical range: How so many species co-exist in the tropics. The American Naturalist. 1989;**133**:240-256

[102] Brook BW, Traill LW, Bradshaw CJA. Minimum viable population sizes and global extinction risk are unrelated. Ecology Letters. 2006;**9**(4):375-382

[103] Malcolm JR, Liu C, Neilson RP, Hansen L, Hannah L. Global warming and extinctions of endemic species from biodiversity hotspots. Conservation Biology. 2006;**20**(2):538-548

[104] Bond WJ, Parr CL. Beyond the forest edge: Ecology, diversity and conservation of the grassy biomes. Biological Conservation [Internet]. 2010;**143**(10):2395-2404. Available from: http://dx.doi.org/10.1016/j.biocon.2009.12.012

[105] Mittermeier RA, Myers N, Thomsen JB, da Fonseca GAB, Olivieri S. Biodiversity hotspots and major tropical wilderness areas: Approaches to setting conservation priorities. Conservation Biology. 1998;**12**:516-520

[106] Overbeck GE, Pfadenhauer J. Adaptive strategies in burned subtropical grassland in Southern Brazil. Flora. 2007;**202**:27-49

[107] Uys RG. Patterns of plant diversity and their management across South African rangelands. Doctoral Thesis. University of Cape Town; 2006

[108] Stott P. Stability and stress in the savanna forests of mainland south-east Asia. Journal of Biogeography. 1990;**17**:373-383

[109] Sankaran M. Diversity patterns in savanna grassland communities: Implications for conservation strategies in a biodiversity hotspot. Biodiversity and Conservation. 2009;**18**:1099-1115

[110] Fisher BL, Robertson HG. Comparison and origin of forest and grassland ant assemblages in the high plateau of Madagascar (Hymenoptera: Formicidae). Biotropica [Internet]. 2002;**34**:155-167. Available from: http://research.calacademy.org/research/entomology/personnel/CVs/pdfs/ComparisonandOrigin2002.pdf

[111] Bond, WJ, Silander, JA, Ranaivonasy, J, Ratsirarson J The antiquity of Madagascar's grasslands and the rise of C4 grassy biomes. Journal of Biogeography. 2008;**35**:1743-1758

[112] Parr, CL, Andersen, AN, Chastagnol, C, Duffaud C. Savanna fires increase rates and distances of seed dispersal by ants. Oecologia. 2007;**151**:33-41

[113] Skowno AL, Bond WJ. Bird community composition in an actively managed savanna reserve importance of vegetation structure and vegetation composition. Biodiversity and Conservation. 2003;**12**:2279-2294

[114] Strayer DL, Power ME, Fagan WF, Pickett STA, Belnap J. A classification of ecological boundaries. Bioscience. 2003;**53**(8):723-729

[115] Gaston KJ, Rodrigues ASL, van Rensburg BJ, Koleff P, Chown SL. Complementary representation and zones of ecological transition. Ecology Letters. 2001;**4**(1):4-9

[116] Kark S. Effects of ecotones on biodiversity. Levin SA, editor. In: Encyclopedia of Biodiversity (Second Edition) [Internet]. 2nd ed. Oxford: Elsevier; 2013. pp. 142-148. Available from: http://www.sciencedirect.com/science/article/pii/B9780123847195003014

[117] McNeely JA, Miller KR, Reid WV, Mittermeier RA, Werner TB. Conserving the World's Biological Diversity. Washington, D.C: International Union for Conservation of Nature and Natural Reources, World Resources Institute, Conservation International, World Wildlife Fund-US, and the World Bank, Gland, Switzerland, and Washington, D.C.; 1990

[118] Myers N. The biodiversity challenge: Expanded hot-spots analysis. Environmentalist [Internet]. 1990;**10**(4):243-256. Available from: http://dx.doi.org/10.1007/BF02239720

[119] Bibby C, Collar N, Crosby M, Heath M, Imboden C. Putting Biodiversity on the Map: Priority Areas for Global Conservation. Review. BirdLife International; 1992. 90 p. Published by the American Ornithological Society. The Auk. 1993;**110**(2):423-424

[120] Brooks D, McLennan D. Historical ecology: Examining phylogenetic components of community evolution. Ricklefs R, Schluter D, editors. In: Species Diversity in Ecological Communities. Chicago, IL: Univ. Chicago Press; 1993. pp. 267-80

[121] Erwin D. The Great Paleozoic Crisis. New York: Columbia Univ. Press; 1993. 327 p

[122] Hampe A, Petit RJ. Conserving biodiversity under climate change: The rear edge matters. Ecology Letters. 2005;**8**(5):461-467

[123] Sandel B, Arge L, Dalsgaard B, Davies RG, Gaston KJ, Sutherland WJ, et al. The influence of late quaternary climate-change velocity on species endemism. Science [Internet]. 2011;**334**(November):660-664. Available from: http://www.ncbi.nlm.nih.gov/pubmed/21979937

[124] Chritz KL, Blumenthal SA, Cerling TE, Klingel H. Hippopotamus (H. amphibius) diet change indicates herbaceous plant encroachment following megaherbivore population collapse. Scientific Reports [Internet]. 2016;**6**(February):32807. Available from: http://www.nature.com/articles/srep32807

Chapter 4

A Synopsis of Global Mapping of Freshwater Habitats and Biodiversity: Implications for Conservation[1]

Ryan A. McManamay, Natalie A. Griffiths,
Christoper R. DeRolph and Brenda M. Pracheil

Additional information is available at the end of the chapter

http://dx.doi.org/10.5772/intechopen.70296

Abstract

Accurately mapping freshwater habitats and biodiversity at high-resolutions across the globe is essential for assessing the vulnerability and threats to freshwater organisms and prioritizing conservation efforts. Since the 2000s, extensive efforts have been devoted to mapping global freshwater habitats (rivers, lakes, and wetlands), the spatial representation of which has changed dramatically over time with new geospatial data products and improved remote sensing technologies. Some of these mapping efforts, however, are still coarse representations of actual conditions. Likewise, the resolution and scope of global freshwater biodiversity compilation efforts have also increased, but are yet to mirror the spatial resolution and fidelity of mapped freshwater environments. In our synopsis, we find that efforts to map freshwater habitats have been conducted independently of those for freshwater biodiversity; subsequently, there is little congruence in the spatial representation and resolution of the two efforts. We suggest that global species distribution models are needed to fill this information gap; however, limiting data on habitat characteristics at scales that complement freshwater habitats has prohibited global high-resolution biogeography efforts. Emerging research trends, such as mapping habitat alteration in freshwater ecosystems and trait biogeography, show great promise in mechanistically linking global anthropogenic stressors to freshwater biodiversity decline and extinction risk.

Keywords: ecology, streams, rivers, lakes, wetlands, fish, crafyish, mussels, amphibians

[1]This manuscript has been authored by UT-Battelle, LLC under Contract No. DE-AC05-00OR22725 with the U.S. Department of Energy. The United States Government retains and the publisher, by accepting the article for publication, acknowledges that the United States Government retains a non-exclusive, paid-up, irrevocable, worldwide license to publish or reproduce the published form of this manuscript, or allow others to do so, for United States Government purposes. The Department of Energy will provide public access to these results of federally sponsored research in accordance with the DOE Public Access Plan (http://energy.gov/downloads/doe-public-access-plan).

1. Introduction

Our knowledge of Earth's ecosystems and biodiversity is growing at rates that exceed our ability to accurately predict regional species pools [1]. Recent estimates of Earth's biodiversity suggest that the planet boasts a total of 8.7 million species, 87% of which are yet to be described [2]. Yet while our comprehension of the magnitude and appreciation of species diversity grows, many have suggested we are currently within the Earth's six mass extinction event [3, 4], in which rates of species loss are unprecedented compared to past extinction events. Indeed, cataloguing biodiversity is a catalyst for global conservation efforts. The International Union for the Conservation of Nature (IUCN) has assessed over 77,300 species, of which 29,530 (38%) are classified as threatened, endangered, or critically endangered, and >10,000 more (13%) species listed as vulnerable [5]. While only 0.01% of Earth's surface water occurs in rivers, lakes, and swamps, >126,000 (7%) of the Earth's described species are found in freshwaters [6, 7]. Therefore, freshwater species especially are in serious jeopardy of extinction.

Dudgeon et al.'s [6] review of threats and conservation challenges to global freshwater biodiversity came at a much-needed time and addressed information gaps limiting our knowledge of these systems. The authors suggested (correctly) that there was no global comprehensive analysis of freshwater biodiversity comparable to those conducted for terrestrial systems [8]. Additionally, there was no comprehensive mapping of inland waters. The lack of this information prohibited our collective ability to inform large-scale conservation and prioritizing species and habitat protection. Since that time, many have answered the call to map global freshwater habitats and biodiversity to inform large-scale conservation. Just 2 years later, in 2008, the first seamless high-resolution map of global river hydrography was developed [9], and the first global biogeographical regionalization of freshwater biodiversity was completed [10].

In more recent years, significant advances in mapping aquatic habitats—specifically rivers, lakes, and wetlands—have been made at the global scale (e.g., [11–13]). Much of the progress in spatially depicting freshwater ecosystems has been the result of new globally comprehensive remote sensing technologies [13], but also significant efforts by scientists to collate disparate data sources [14]. As new datasets and geospatial products emerge with increasing spatial resolution, estimates of the spatial extent and importance of freshwater ecosystems in global biogeochemical cycles have also increased [15–17]. While efforts to develop comprehensive inventories and maps of the distribution of the world's freshwater fauna have dramatically increased [18, 19], these efforts have remained separate from those of freshwater habitat mapping.

Herein, we briefly review the status and recent history of global mapping of freshwater habitats, their biodiversity, and human disturbances. First, we provide an overview of the efforts and datasets to empirically map rivers, lakes, reservoirs, and wetlands at the global scale, and compare these to theoretical estimates of the spatial coverage of unobserved features. This provides an assessment of the accuracy and comprehensiveness of global freshwater habitat mapping. Secondly, we discuss the current state of global freshwater biodiversity mapping and provide sources of information and various approaches used. We compare the spatial scales and resolution of biodiversity and freshwater habitat mapping to identify potential overlap and information gaps. Additionally, we discuss various approaches to map the global extent of human disturbances in freshwater systems. Finally, we discuss emerging themes, but also gaps and research needs for continuing to improve our knowledge of patterns in freshwater species and their habitats. We also present summaries of the various databases used in supporting these efforts, which to our knowledge have not been previously summarized in one publication.

2. Global freshwater habitat mapping efforts

Global estimates of freshwater ecosystem coverages have been developed through both theoretical [20] or empirical means [21], or a combination of both [11]. Theoretical constructs, for example, might assume relationships between the size, distribution, and bifurcation of rivers (i.e., network theory) to quantify size and distribution of rivers within a region [20]. Likewise, theoretical relationships of size versus distribution are commonly used to estimate the frequency and size of unobserved waterbodies [22]. In contrast, empirical estimates typically rely on spatial observations from remote sensing data. Because the geospatial representation of waterbodies is limited to the spatial fidelity of mapping efforts, the number and areas of waterbodies provided through empirical observation is consistently smaller than that estimated theoretically. This comparison is important, however, in that it yields insights into the current state (i.e., comprehensiveness and granularity) of global freshwater mapping efforts. In the following sections, we review and compare approaches to obtaining global scale estimates of three different freshwater ecosystem types: rivers and streams, lakes and reservoirs, and wetlands. Estimation methods and datasets vary for each of these aquatic ecosystem types and influence their respective global estimates. We also devote particular attention to trends in freshwater mapping efforts within the United States.

2.1. River and streams

Global estimates of river and stream mileage and area range widely, with aerial estimates provided more frequently than distances. The latest and largest estimates of river length and area are over 88.3 million km and 662,100 km^2, respectively [20]. To provide these estimates, Downing et al. [20] used two approaches, one reliant on stream network theory and empirical data on stream widths and the other estimating the fraction of continental area occupied by streams while correcting for the unresolved small stream portion. The authors first estimated global river number, length, and area according to stream order by relying on relying on river geometry and scaling laws [23, 24] and known bifurcation ratios and stream length-order equations [25]. Stream widths among different order streams were obtained from literature or aerial imagery and applied to the number and lengths of streams. In the second method, estimates of the fraction of river area per land for well-studied landscapes were extrapolated to the global land area, which led to a very close second approximation, 640,400 km^2.

Empirical estimates of global river length and area from mapping efforts are far less than the maximum theoretical estimates [20]. The Digital Chart of the World (DCW) estimates global stream length at 16.6 million km [26, 27]. HydroSheds (basins and stream networks) were developed from global digital elevation models (DEMs) which increased the estimate to 27.3 million km (derived from 15 arc-second resolution) (**Figure 1**) [9]. The Hydro1K database is currently the highest resolution empirical estimates of global stream length [28], which constitutes 53% of the highest theoretical estimates [20]. Previous estimates of global river area range from 360,000 to 510,000 km^2 (**Table 1**). The Global Lakes and Wetlands Database (GLWD) is a compilation of at least 17 different datasets of regional to global registers, inventories, and digital maps according to different spatial extents [21]. Their estimate of 360,000 km^2 of global river area was dependent upon aerial and satellite imagery of >5th order rivers and streams [20].

The spatial distribution and quantification of global river and stream mileage is limited to the resolution of widespread DEMs and, in turn, derived stream networks [31, 32]. Increased spatial resolution [33] and new algorithms for deriving stream networks [31] have continually increased the accuracy of spatial representations of global rivers (**Figures 1** and **2**). The finest resolution of consistent global-extent elevation grids is >90 m [9, 28], which will grossly underrepresent small stream systems. According to the DCW, the length of streams and rivers within the conterminous-US (CONUS) totals 727,326 km (almost 29,000 reaches) whereas the HydroSheds database (15 arc-second) estimates the same distance as almost 1.9 million km (238,405 reaches) (**Figure 3**). In contrast, the total mileage is 5.7 million km (2.98 million reaches) according to the NHD plus medium resolution dataset (1:100k scale) [34], which was constructed on the basis of 30-m DEM resolution [35]. The NHD High-Resolution Dataset (1:24k scale), however, estimates stream length for the CONUS at 1.2 million km (**Figure 3**) [36]. While mapping perennial systems seems straightforward, accurately mapping ephemeral systems from flow accumulation thresholds is difficult. Even the NHDplus dataset under-represents the small headwater systems apparent in the high-resolution National Hydrography Dataset (1:24k scale), which also under-represents potential ephemeral systems (**Figure 2**).

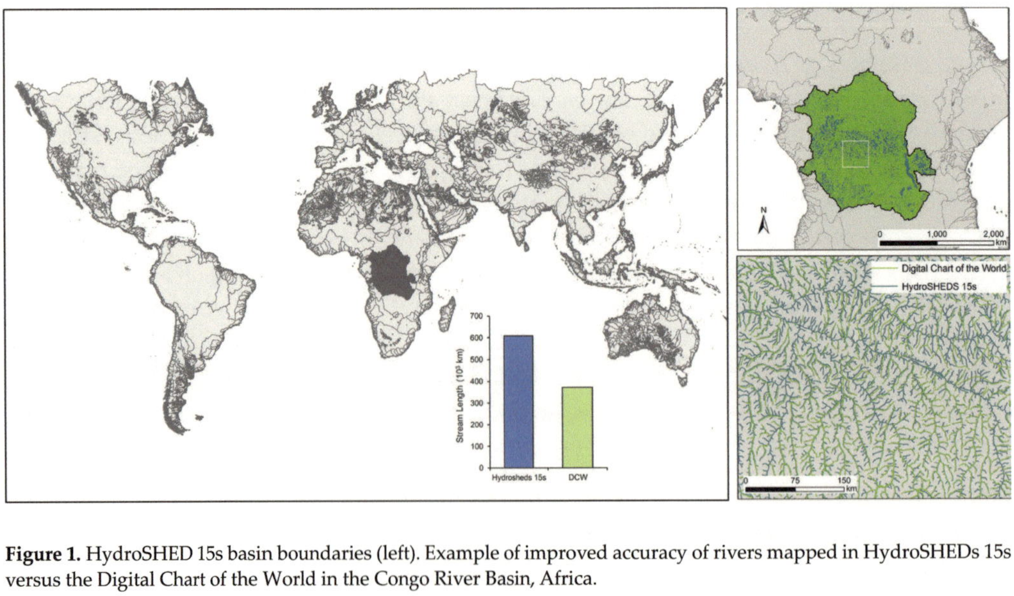

Figure 1. HydroSHED 15s basin boundaries (left). Example of improved accuracy of rivers mapped in HydroSHEDs 15s versus the Digital Chart of the World in the Congo River Basin, Africa.

http://dx.doi.org/10.5772/intechopen.70296

Study or database	Length (km)	Area (km²)
Theoretical		
Downing et al. [20]: A	88,325,340	662,100
Downing et al. [20]: B		640,400
Downing et al. [20]: C		485,000
Aufdenkampe et al. [29]		510,000
Downing [30]		508,000
Empirical		
HydroSheds [9]	27,300,269	
Global Wetlands and Lakes Database [21]		360,000
Digital Chart of the World [26, 27]	16,610,004	
Hydro1K [28]	46,900,425	

Downing et al. [20] use three different approaches to estimating stream and river area as denoted by A, B, and C (see text).

Table 1 Theoretical and empirical estimates of global stream and river length and area provided by different studies and datasets.

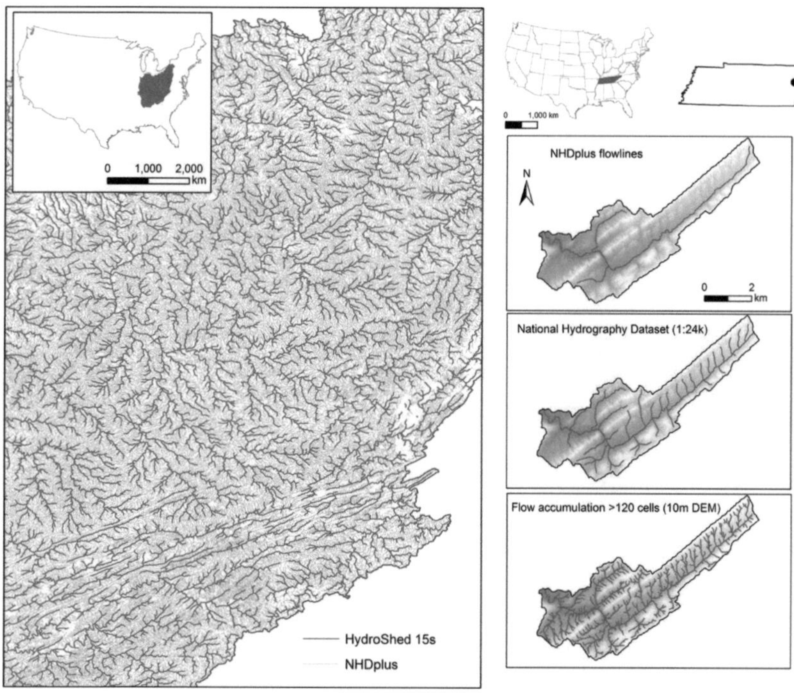

Figure 2. Comparison of HydroSHEDs to NHDPlus (1:100k) flowlines in the Ohio and Tennessee River Basins of the US (left). Example of the increased spatial resolution provided by the National Hydrography Dataset (High-resolution, 1:24k) over that of NHDPlus in Bear Creek, near Oak Ridge, Tennessee, USA. However, ephemeral channels are likely even underestimated by the NHD High-resolution dataset.

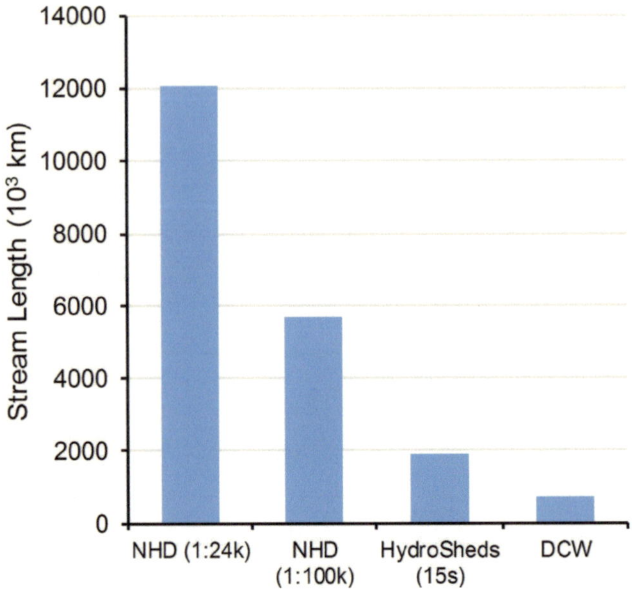

Figure 3. Total continental US stream distance represented by four spatial datasets depicting river networks.

Interestingly, global length-stream order relationships do not follow global area-stream order relationships. For example, the number and length of 1st order systems in the world are, by far, numerically dominant constituting 52% of global river length (28.5 million and 45.7 million km^2, respectively) [20]. However, global river area is dominated by larger order systems (≥6th order), which represent 65% of total river area. Size-specific stream distribution estimates are extremely important for accurately portraying or modeling the distribution of aquatic organisms.

2.2. Lakes, reservoirs, and farm ponds

Studies estimating the global extent of lakes and reservoirs were more numerous than those estimating river and stream distributions. Global numbers of lakes range from 800,000 to 304 million whereas cumulative area of world lakes ranges from 2.3 to 5 million km^2 (**Table 2**, **Figure 4**). Human construction of reservoirs has been extensive, the most current estimate at 16.7 million waterbodies with a cumulative surface of 305,723 km^2, an area equivalent to increasing the world's naturally occurring terrestrial water surface by 7.3% [11]. Other estimates of global reservoir surface area range from 150,000 to 600,000 km^2, depending on the source and whether regulated natural lakes are included (**Table 2**). Only one study provided an estimate of global farm pond coverage (77,000 km^2) using relationships between the fraction of farm pond area within farm land and annual precipitation [22].

Similar to rivers and streams, lakes and reservoirs have been estimated using both empirical observation of available geospatial datasets or via extrapolation of observed data to unobserved features. Until recently, theoretical estimates of lakes exceeded that of empirically derived estimates. New high-resolution satellite imagery provided means to observe lakes

Area	Lakes	Reservoirs	Farm ponds
	10^3 km^2	10^3 km^2	10^3 km^2
Kelly et al. (1994) [37]		500	
Pearce (1996) [38]		600	
Meybeck (1995) [39]	2300–2600		
Lehner and Doll (2004) [21]*	2428	251	
Lehner and Doll (2004) [21]*	3200		
McDonald (2012) [40]	3800		
Downing et al. [22]	4200	260	77
St. Louis et al. (2000) [41]		150	
Lehner et al. (2011) [11]*		305	
Messager et al. (2016) [42]*	2677	250	
Verpoorter et al. (2014) [13]*	5000		
Number	10^3	10^3	10^3
Meybeck (1995) [39]	800–1300		
Lehner and Doll (2004) [21]*	246	0.822	
Lehner and Doll (2004) [21]*	15100		
McDonald et al. (2012) [40]	64000		
Downing et al. [22]	304000		
Lehner et al. [11]*		16700	
Messager et al. [42]*	1421	7	

*Empirical estimates.

Table 2 Global estimates of the area and number of lakes, reservoirs, and farm ponds according to different studies.

>0.002 km^2 [13]. Using this technology, the GLObal WAter BOdies database (GLOWABO) was developed for 117 million lakes with a total surface area of 5 million km^2 [13]. This surface area estimate exceeds that of the highest theoretical estimate [20], but is still smaller in total lake abundance (**Figure 4**).

The development of reservoir mapping datasets has provided valuable spatial representations of waterbodies in recent years. For example, the GLWD dataset consists of polygon shapefiles of approximately 250,000 lakes and reservoirs >0.1 km^2 and raster datasets of other lakes, reservoirs, and wetland coverages [21]. The GLWD included only information for the world's largest reservoirs (storage >0.5 km^3) either because spatial information was limiting or existing lake datasets did not explicitly clarify whether a given waterbody was manmade. Because of the incomplete nature of global datasets on impoundments, the Global Reservoir and Dam database (GranD) was developed as a compilation of spatial coverages of 6862 reservoir polygons and associated dams and attributes [11]. More recently, a new geospatial coverage of

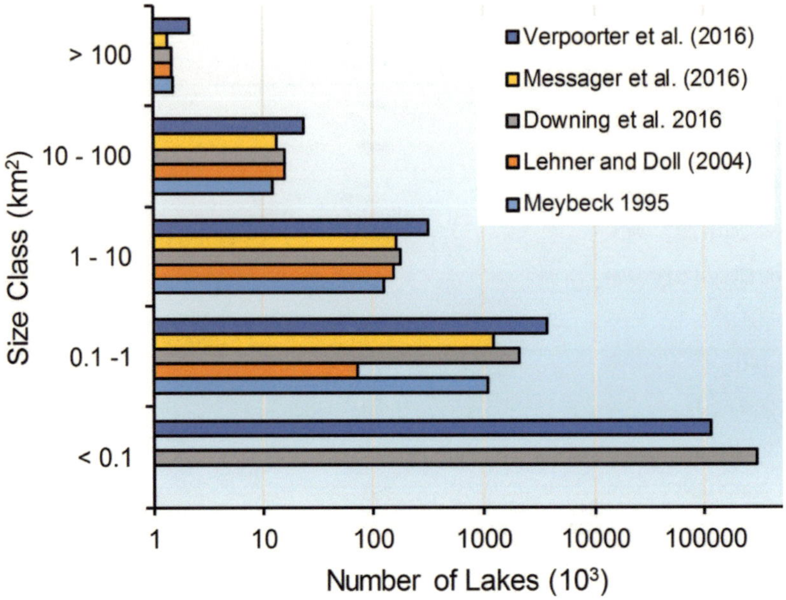

Figure 4. Global lake abundance estimated by several different studies.

global lakes and reservoirs, HydroLakes, was developed and includes hydrologic attributes, such as volume and residence time, using a geo-statistical model [42] (**Figure 5**). Within the US, the NHDplus (1:100k) dataset provides coverage of lakes and areas as polygons, an area estimated at almost 250,000 km^2; however, this dataset misses small waterbodies, especially farm ponds. The NHD high-resolution (1:24k) dataset estimates lake and reservoir area coverage as approximately 890,000 km^2, almost 3.5 times higher than that of NHDplus.

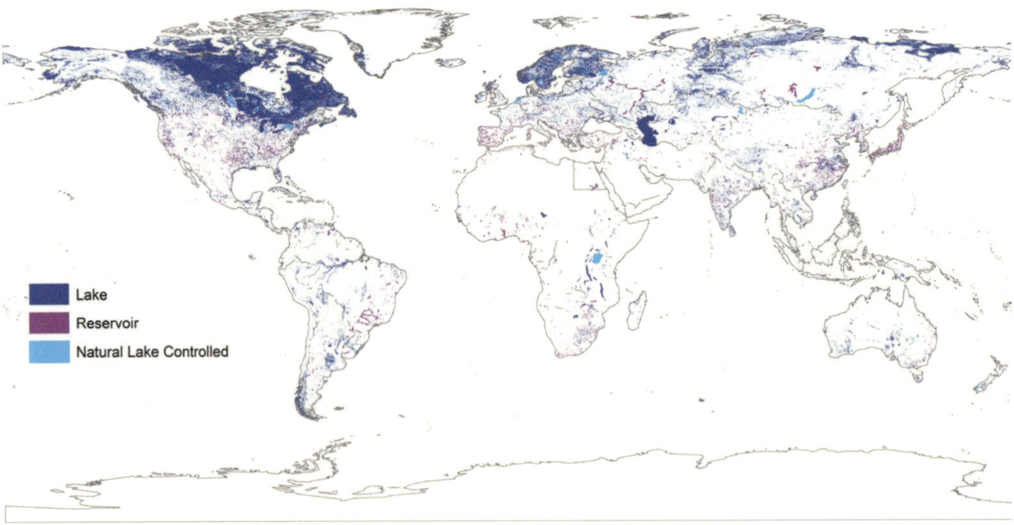

Figure 5. HydroLakes database depiction of global lakes and reservoirs.

The most numerous lake and reservoir waterbodies are very small (<0.1 km^2) (**Figure 4**), yet these are typically omitted from most maps (with recent exceptions, [13]). To estimate the size and distribution of these smaller waterbodies, Pareto distributions of log-abundance versus log-size are fit to observed larger lakes and then those coefficients are used to extrapolate the abundance of smaller, unobserved lakes [43] or reservoirs [11]. Obviously, these estimates do not come without error, with some suggesting that numbers of small lakes and any related scaling estimates (e.g., carbon fluxes) are unreliable [44].

2.3. Wetlands

Wetlands are transitional systems by nature, making them difficult to distinguish from other waterbodies. A distinction is provided by the U.S. Fish and Wildlife Service (USFWS) [45], which defines wetlands as "lands transitional between terrestrial and aquatic systems where the water table is usually at or near the surface or the land is covered by shallow water". USFWS [45] goes on to list three main attributes of wetlands: "(1) at least periodically, the land supports predominantly hydrophytes, (2) the substrate is predominantly undrained hydric soil, and (3) the substrate is non soil and is saturated with water or covered by shallow water at some time during the growing season of each year." In contrast, the International Union for the Conservation of Nature (IUCN) broadens the definition of wetlands to be all-inclusive of "areas of marsh, fen, peatland or water, whether natural or artificial, permanent or temporary, with water that is static or flowing, fresh, brackish or salt, including areas of marine water the depth of which at low tide does not exceed 6 m" [46]. For our purposes, we include wetlands as any waterbody or part of the landscape that falls within the definitions above, but cannot be distinguished as a lake, reservoir, pond, river or stream.

Unfortunately, there is little consistency in the nomenclature distinguishing among various waterbodies in the spatial datasets used to estimate global coverage of wetlands. The GLWD is commonly used in representations of wetlands across the globe (**Figure 6**). Many of the spatial datasets contributing to the GLWD, however, have contrasting naming conventions for waterbodies [21]. In particular, the DCW does not distinguish between vectors portraying lakes, reservoirs, larger rivers, and wetlands [26]. In comparison, the Wetlands Map of the World Conservation Monitoring Center (WCMC) includes 20,000 wetland and lake polygons classified into 21 types and represents the most comprehensive and accurate vector map of the world's wetlands [47]. As opposed to representing wetlands as vectors or polygons, other mapping efforts display wetlands as raster maps. For example, the US Geological Survey Global Land Cover Characteristics (GLCC) database [48] and MODerate resolution imaging spectroradiometer (MODIS) data [49] provides classification of global landcover, including wetlands, as 30 second grids (MODIS). Others have developed global wetland land cover maps at coarser resolutions using varied methodologies [50–52]. Because of the uncertainties on global wetland extents and inventories, the Ramsar Convention on Wetlands has promoted new efforts and advanced remote sensing technologies to provide new and improved global wetland inventories [53, 54].

Similar to other freshwater systems, estimates of the global coverage of wetlands have increased over time with advances in higher-resolution spatially comprehensive datasets.

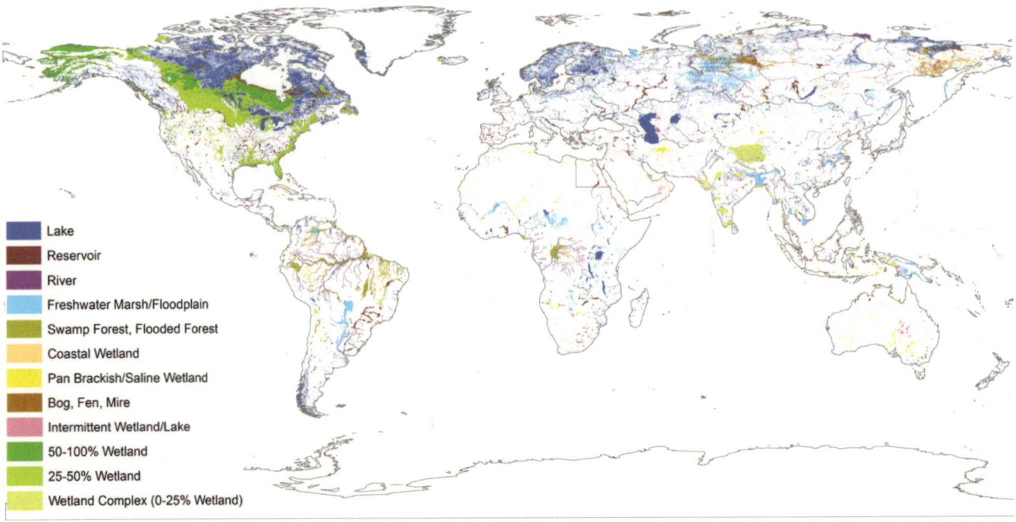

Figure 6. Map of global waterbodies based on the Global Lakes and wetlands database (GLWD).

Early estimates (pre-2000) ranged from 4.3 to 5.3 million km^2 whereas current estimates approach almost 13 million km^2 (**Table 3**). However, the highest estimate may be an overestimate inclusive of lake and reservoir waterbodies [57] relative to the reference [21] estimate of 9.2 million km^2. Within the US, wetlands are depicted by a few vector and raster datasets. For the conterminous US, the Multi-Resolution Land Characteristics Consortium (MRLC) provides National Land Cover Databases (NLCD) as raster images [58]. According to the 2011 NLCD data, the area classified as woody or herbaceous wetlands sums to 417,442 km^2. Open water constitutes almost the same spatial area, 422,111 km^2. The USFWS maintains the

Study	Wetlands (10^3 km^2)
Lehner and Doll [21]	9167
Williams [55]	8558
Mitch and Gosselink [56]	7000 - 9000
Mathews and Fung [50]	5260
Cogley [51]	4340
Sillwell-Soller et al. [52]	4795
GLCC [48]	1093
MODIS [49]	1291
Gross Wetlands Map [21]	11711
Finlayson et al. [57]	12800

Numbers provided by Lehner and Doll [21].

Table 3 Global areal estimates of wetland coverages according to different studies.

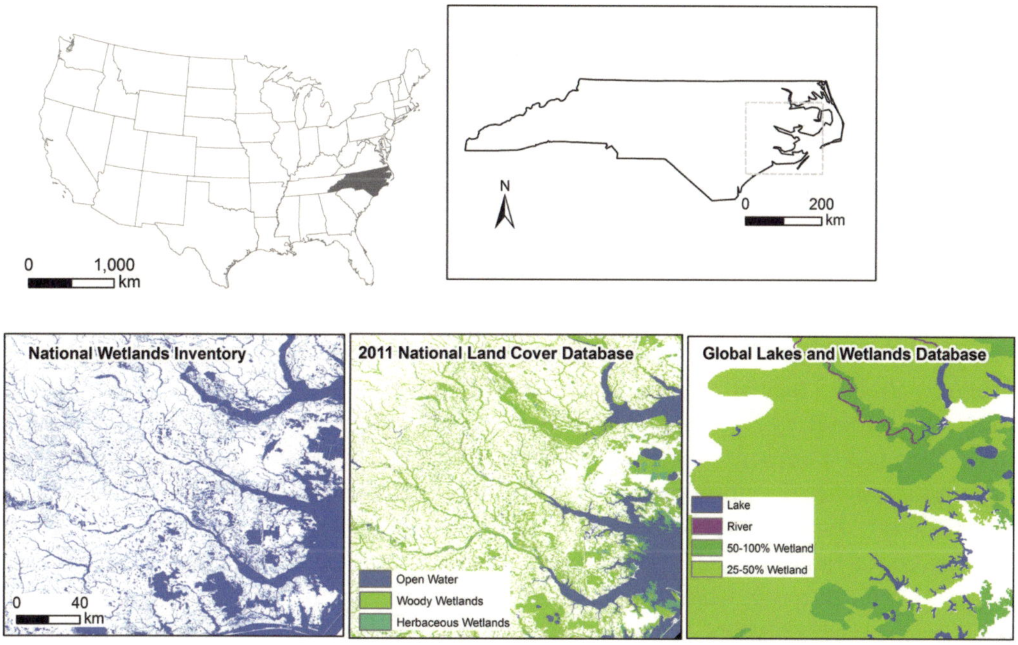

Figure 7. Comparison of wetland maps derived from the National Wetlands Inventory (NWI), the National Land Cover Database (NLCD), and the Global Lakes and Wetlands Database (GLWD) for a coastal portion of the State of North Carolina located in the eastern United States. Examples of types of wetland databases available in the conterminous US.

National Wetland Inventory (NWI), a database of polygons and associated very detailed classification framework for the conterminous US [59] (**Figure 7**). The NWI provides a status update of the nation's wetlands every five years with the latest 2009 report indicating there were 445,559 km^2 of wetlands, 95% of which are freshwater systems [60]. The difference of 28,118 km^2 between NWI and NLCD estimates of wetland area for the entire conterminous US suggests differences in the approaches taken to classify wetlands (**Figure 7**). Both of these datasets, however, far exceed the spatial granularity of wetlands depicted by the GLWD (**Figure 7**).

3. Global biodiversity mapping efforts

Global and continental-scale mapping of freshwater species distributions has lagged freshwater habitat mapping efforts in terms of finer spatial granularity. More specifically, there are mismatches between the resolution of current global biogeography efforts and the spatial fidelity of waterbodies in the landscape. This makes intuitive sense for two main reasons: (1) The presence of a species within a given area typically requires in situ observation, as opposed to detection via remote sensing technologies, such as in the case of waterbodies and other landscape features. That being said, remote sensing of biodiversity is a rapidly growing area of research [61], with potential new capabilities for direct aerial observation of biota [62]. (2) Most observations of species are discrete points in space and time, are influenced by methods of detection, and are not spatially comprehensive. Hence, extending species presences into unsampled areas requires various levels of inference ranging from summarization into regions or watersheds to sophistical statistical models predicting probability of presence using a suite of predictor variables characterizing habitat. Obviously, the first approach requires less resources and information, whereas the latter approach requires rich information on descriptions of habitat, not just the features themselves.

3.1. A synopsis of published global biodiversity mapping

Generally, we found little congruence between global mapping of biodiversity and global mapping of freshwater habitats (**Table 4**). Only two studies in **Table 4** used spatial products from recent global habitat mapping efforts [19, 72]. Richman et al. [19] summarized crayfish range maps from IUCN and georeferenced occurrences (from experts) in Hydro1K basins [28] to examine factors responsible for their decline. All but one of the studies outlined in **Table 4** have been published within the last 15 years, and opposite as expected, species mapping efforts do not display a clear trend of increasing spatial granularity over time. In contrast, studies seem to summarize biogeographical information at the coarsest scales sufficient to achieving their purpose, which in most cases, was related to examining declines in species and threats to their existence. Spatial resolutions of freshwater species mapping ranged from biogeographic regions and range estimates (polygons) to 96-km^2 gridded cells and small watersheds (e.g., Hydro1K).

Source	Description	Spatial resolution	Source
Fish			
Oberdorff et al. [63]	Analyze fish species richness patterns across continents and show that species-area and species-energy relationships explain most of the variation	Major drainage basins (n = 292)	Multiple published sources
Amarasinghe & Welcome [64]	Developed models of fish species richness from natural lake characteristics	Nature lake features	Multiple published sources; International Lake Environment Committee Foundation (ILEC) global lake database [65]
Xenopoulos et al. [66]	Use global hydrologic model to simulate scenarios of future fish species loss with losses in river discharge from climate change and withdrawal	Major drainage basins (n = 325)	Oberdorff et al. 1995 [63]; FishBase [67]
Abell et al. [10]	Developed first global biogeographic regionalization of Earth's freshwater systems based on composition of freshwater fish species	Freshwater ecoregions (n = 397)	Multiple

Source	Description	Spatial resolution	Source
Oberdorff et al. [68]	Developed a framework of mechanisms and processes driving global and regional patterns in fish richness	Major drainage basins	Multiple published sources
Liermann et al. [69]	Use spatial distribution of fish, their traits, and current dam development to examine risks of fish species loss	Freshwater ecoregions (n = 397)	Abell et al. 2008 [10]
Bross et al. [70]	Developed a database of native, endemic and non-native fish species richness in major basins of the world	Major drainage basins (n >1000)	Multiple published sources
Toussaint et al. [71]	Examine world patterns in functional diversity of fish relative to species diversity	Biogeographic regions (n = 6)	Bross et al. 2013 [70]
Winemiller et al. [72]	Examined patterns in fish biodiversity and endemic species overlapping with current and proposed dam construction in the Amazon, Congo, and Mekong River basins	Freshwater ecoregions; hydroBasins	Abell et al. 2008 [10]; IUCN [73]
Amphibians			
Stuart et al. [74]	Status and trends of worldwide amphibian declines and extinctions. Mapped species distributions by reason for decline	1^0 Cell	Global Amphibian Assessment (IUCN) [75]
Gallant et al. [76]	Global assessment of land use dynamics in the context of amphibian distributions	Global ecoregions (n = 21)	Global Amphibian Assessment (IUCN) [75]
Sodhi et al. [77]	Global analysis to quantify the influences of life history, climate, human density, and habitat loss on declines and extinction of 45% of known amphibians	Range maps	Global Amphibian Assessment (IUCN) [75]
Wake and Vredenburg [3]	Global assessment of the decline and extinction of amphibians	Country	Multiple
Rödder et al. [78]	Global risk assessment for amphibian extinction for the Panzootic Chytrid Fungus	0.5^0 Cell	Global Amphibian Assessment (IUCN) [75]

Source	Description	Spatial resolution	Source
Hof et al. [79]	Assess the current and future interactions of climate change, land-use change, and spread of the pathogenic fungal disease chytridiomycosis on amphibian species declines	2^0 Cell	Multiple
Ficetola et al. [80]	Assessment of error in global range maps for amphibians	Range maps; point distributions	Global Amphibian Assessment (IUCN) [75]; GBIF [81]; Check List Online Journal [82]
Mussels			
Graf and Cummings [83]	Review of systematics and global diversity of freshwater mussel species	Geographic regions (n = 32)	MUSSEL Project [84]
Nobles and Zhang [85]	Assessment of global biodiversity loss in mussels including threats and solutions	Biogeographic regions (n = 6)	Multiple published sources
Crayfish			
Crandall and Buhay [86]	Description of global diversity in crayfish	Continents	Multiple
Richman et al. [19]	Evaluation of factors responsible for global declines in crayfish	HydroIK river basins	IUCN; expert georeference collection efforts
Multiple taxa			
Rodrigues et al. [87]	Examination of global protected areas in representing species diversity (includes amphibians, mammals, birds, turtles).	0.5^0 Cell	IUCN [73]
Rodrigues et al. [88]	Global gap analysis assessing the extent of protected land coverage for representation of biodiversity including amphibians, mammals, freshwater turtles and tortoises, and globally threatened birds	0.25^0 Cell	IUCN [73]
Grenyer et al. [89]	Examine congruence and commonalities in biodiversity and rare and threatened species among amphibians, mammals, and birds	96.3 km^2 grids	Multiple

Source	Description	Spatial resolution	Source
McGeoch et al. [90]	Development of indicators describing relationships between the extent of biological invasion by alien species, its impact on biodiversity and policy response. Species included mammals, birds, amphibians, freshwater fish, vascular plants and marine organisms (including algae, corals, invertebrates and fish)	Countries	Convention on Biological Diversity [117]
Collen et al. [18]	Examined geographical ranges of 7083 freshwater species of mammals, amphibians, reptiles, fishes, crabs and crayfish to examine commonalities in distribution of richness, threatened species, endemism, and congruence in diversity measures among taxa	1^0 Cell	IUCN [73]
Prim et al. [91]	Review of global species biodiversity, their rates of extinction, distribution, and protection (includes amphibians, fish, terrestrial birds, terrestrial mammals, and plants)	Varied (amphibians, 0.5^0 cell; fish, Freshwater Ecoregions, n = 397)	IUCN [73]; Abell et al. 2008 [10]
Jenkins et al. [92]	Assessed the US protected areas with respect to biodiversity of freshwater fish, terrestrial vertebrates, and trees	Varied	Nature Serve [93]; BirdLife International [94]; IUCN [73]; US Geological Survey Tree Database [95]

Table 4. Examples of studies developing or utilizing global freshwater biogeography databases.

In most cases, global mapping of biodiversity has been achieved by summarizing occurrence or estimated range information into spatial units as opposed to developing predictive species distribution models (SDMs) (**Table 4**). There are, however, several global-scale species modeling efforts, many of which are provided as interactive online resources (see following sections). Of freshwater taxa, amphibians and fish mapping efforts have been documented more than crayfish and mussels (**Table 4**), possibly because more vertebrate species have been described and more is known about the details of their life histories, habitat requirements, and conservation status. Additionally, global mapping efforts for amphibians are more common because of the wealth of data for that taxa. In particular, the Global Amphibian Assessment conducted by the International Union for the Conservation of Nature (IUCN) produced polygon range maps for >6000 known amphibian species [75] (**Figure 8**) and was used in six different studies (**Table 4**). The IUCN provides similar spatial data for mammals, reptiles, and marine and freshwater taxa [73]. The range maps are many times converted to gridded raster datasets [74] (**Figure 8**) or overlapped with region polygons to provide summaries of species within those areas (e.g., [76]).

The IUCN recently produced a set of higher-resolution global maps of ranges of freshwater taxa (IUCN) within HydroBasins (240,000 basins globally) [12] (**Figure 9**). One study relied on this resource to examine spatial relationships between fish biodiversity and planned hydropower dam construction in three large basins of the world [72]. The authors suggested that site selection for dams not be conducted purely on the grounds of energy, but should be conducted strategically through tradeoff analyses to conserve the most biodiversity while financing new dams. The IUCN data is currently the best openly available global information on freshwater species occurrences, but has many gaps in spatial coverage (e.g., **Figure 9**). While the Congo and Mekong River (China) basins had sufficient information at the resolution of HydroBasins, the Amazon Basin did not have comprehensive biodiversity mapping at that resolution; hence, reference [72] relied on biodiversity estimates in Freshwater Ecoregions [10], a far coarser alternative. The Amazon basin is over 7 million km^2 yet only contains 13 Freshwater Ecoregions. Obviously, for conservation purposes, higher-resolution granularity is required to inform dam site selection in many areas of the globe. To compensate for lack of knowledge in many areas of the world, other mapping efforts have relied on published resources to compile freshwater species lists within regions or basins [63, 70]. While these resources can fill in important knowledge gaps, they are coarse (presented at the resolution of large basins) and leave large regions of the globe vacant of information (**Figure 9**).

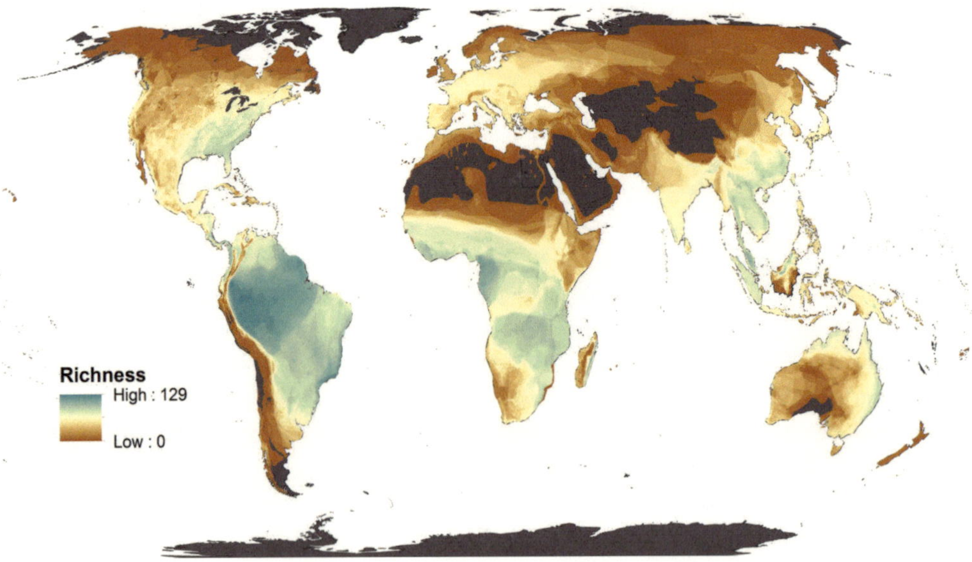

Figure 8. Global amphibian richness from the International Union for the Conservation of Nature (IUCN) Global Amphibian Assessment.

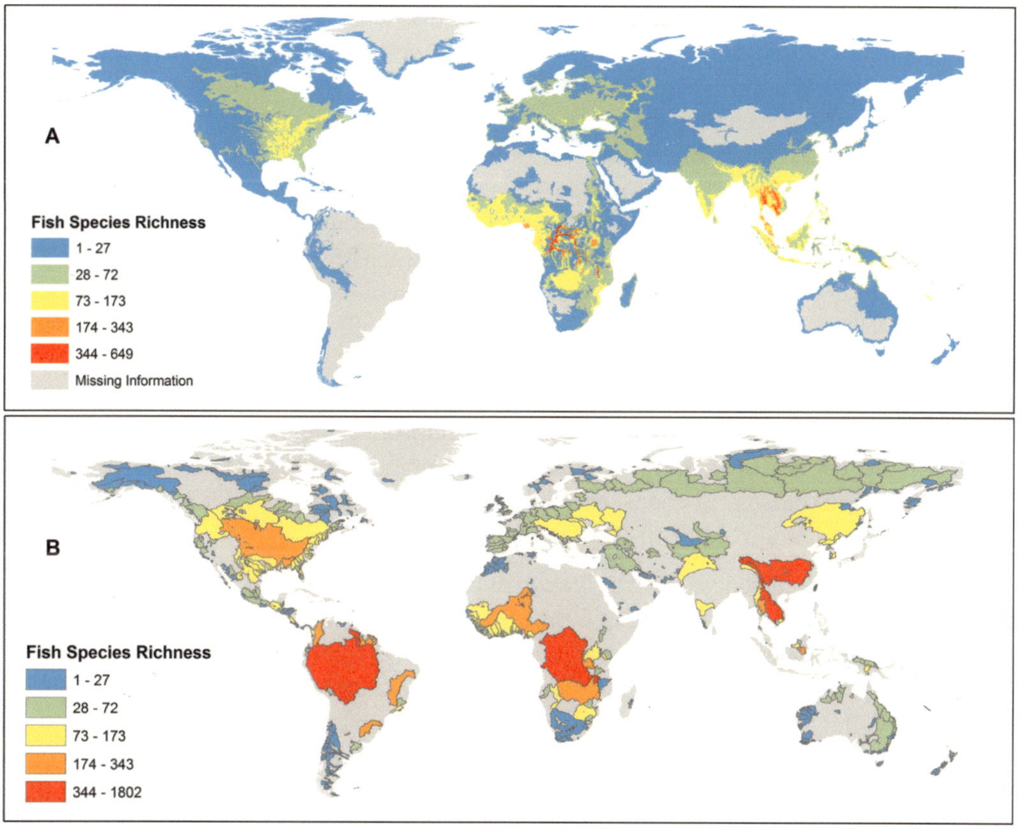

Figure 9. Global maps of fish richness provided by the IUCN [73] and Bross [70].

3.2. What is limiting global high-resolution freshwater species distribution models (SDMs)?

Although many of the world's freshwater species lack formal description, are prone to misidentification, and have few georeferenced occurrences, databases of species observations and species characteristics are growing rapidly. For example, the Global Biodiversity Information Facility (GBIF) currently has over 730 million occurrences for over 1.64 million species and harnesses global community participation [81]. GBIF operates through more formal data publishing, whereas other databases, such as iSPOT [96] provides a platform for crowd-sourced species observations. Additionally, rich databases on species ecology and conservation status have emerged to assist with linking biodiversity with their global freshwater habitat requirements [67, 93]. The wealth of information from georeferenced occurrence databases and descriptive databases suggests that global freshwater biodiversity SDM efforts are not limited by observations, but the inability to extrapolate occurrences to fine-grain freshwater habitats via distribution modeling. This is not to suggest that global freshwater biodiversity SDM efforts are completely absent. Indeed, novel web tools are available to enable users to perform their own SDM projections, both current and future. The Life Mapper project is an online resource that utilizes GBIF observations and global climate, terrain and land cover information to model the current and future distributions of species (including freshwater) [97]. Models of current ranges of species and habitat specifications are calibrated based on existing observations and climate information and used to model future potential ranges based on four climate scenarios spanning 2050 and 2070, according to the International Panel on Climate Change (**Figure 10**). As another example, AquaMaps uses a simplistic "environmental-envelope" method to develop large-scale predictions of marine and freshwater species occurrences [98, 99]. Occurrence data are obtained from GBIF and literature available through FishBase and summarized within bounding basins to constrain subsequent projections of distribution to only natural ranges. Occurrence data are overlain with eight environmental parameters to create an envelope of environmental suitability, which is essentially using the percent of observations (percentiles) in conjunction with local habitat conditions to estimate probability of occurrence [98]. Environmental envelopes are then used to model probabilities of species occurrence based on local conditions. Both the Life Mapper project and Aquamaps are freely available and are a quick approach to developing distribution maps; however, they are still relatively coarse projections, currently set at 10 arc-minutes and 0.5° (30 arc-second) cells, respectively, and do not approximate freshwater habitat features.

We suggest that the current leading limitation of achieving high-resolution global freshwater biodiversity mapping efforts has been a matter of limiting global habitat characteristic data, as opposed to limitations in occurrence data. Even if occurrences for a species are limited, current modeling approaches (e.g., Maxent) are capable of developing SDMs with low sample sizes [100]. By high-resolution, we are referring to the spatial granularity that approximates that of global freshwater habitat features. Recent developments have produced high-resolution depictions of freshwater features in the landscape, but much of these features have little accompanying information on habitat requirements for species (e.g., temperature, hydrology, depth, etc). One exception is a database on world lakes (n = 217) provided by the International Lake Environment Committee Foundation (ILEC), which includes location, morphometric features, climate, water quality, and edaphic variables [65]. This provided an opportunity to model fish species richness in selected natural lakes across the globe [64].

In comparison to terrestrial ecosystems, habitats within freshwater systems are shaped by upstream hydrologic processes, which require sophisticated geospatial summarization methods for appropriate characterization. For example, suppose air temperature is being used as a surrogate of water temperature in a fish species distribution model at the resolution of stream reaches or small watersheds. In this case, air temperature summarized at the location of the individual stream reach is unlikely to be representative of actual water temperature conditions. In contrast, using stream network routing to accumulate air temperature values for the entire upstream drainage network of each reach would be more representative [35]. Until recently, this type of habitat characterization was globally unavailable to support high-resolution freshwater species distributions. A near-global dataset summarizing 324 layers describing climate, land cover, topography, geology, and soils was recently developed for upstream drainage network of HydroSHEDs river reaches [101]. For the US, a comparable dataset is the NHD plus system (1:24K scale), which provides climate, hydrology, and land-use information summarized within the entire upstream network above each stream reach. Many freshwater species distribution modeling efforts have utilized the NHDplus data (1:24k) and architecture

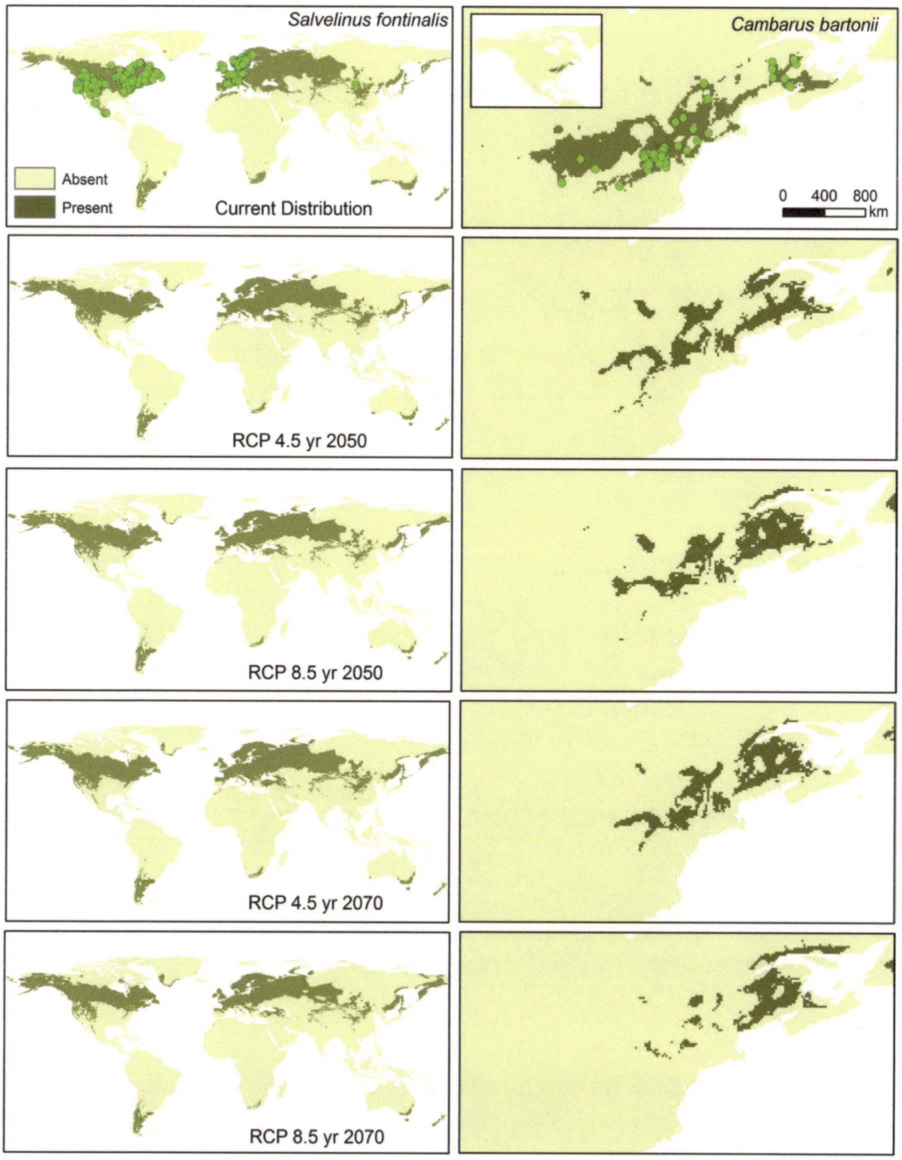

Figure 10. Life map projections of Brook Trout (*Salvelinus fontinalis*) and Appalachian Brook Crayfish (*Cambarus bartonii*) (f-j) distributions for current conditions and future climate projections for 2050 and 2070 under low (4.5 W/m^2) and high (8.5 W/m^2) IPCC representative concentration pathways (RCPs) for radiative forcing levels related to projected greenhouse gas emissions scenarios. Green points represent GBIF occurrences.

because of topological connectivity and habitat predictors offered by the resource [102–107] (**Figure 11**). Although NHDplus is a convenient database to support freshwater species distribution modeling, it does not adequately represent 1st order streams, the majority of which provide habitat for freshwater taxa (**Figure 11**). The NHD High resolution database (1:100k) represents smaller stream systems, but does not provide pre-summarized habitat information.

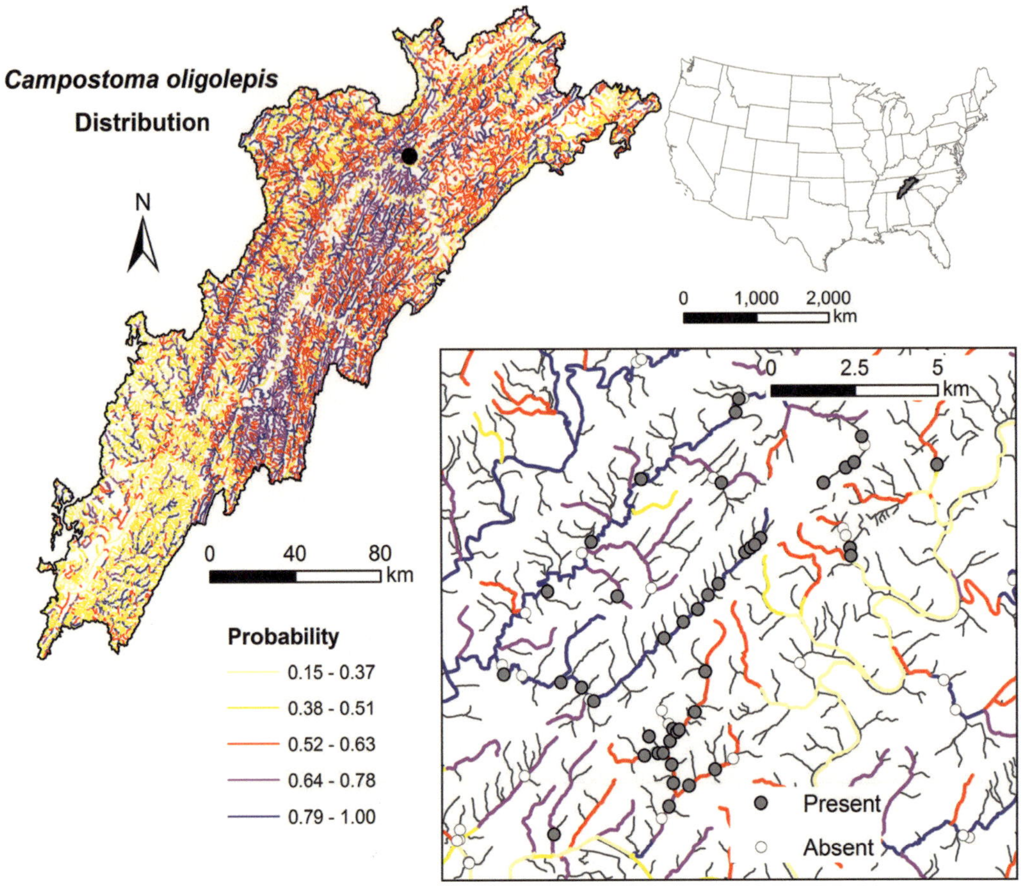

Figure 11. Species distribution model (SDM) developed for Largescale stoneroller (*Campostoma oligolepis*) in the Ridge and Valley and the Southern Appalachian Plateau Ecoregions of the Tennessee River Basin, USA. SDMs are generated for NHDPlus (1:100k) stream reaches and do not account for occurrences in NHD High-resolution stream reaches (smaller gray lines).

For this reason, other studies have developed their own reach datasets with accumulated habitat variables to support freshwater SDMs at resolution comparable to the NHD high-resolution dataset [108].

3.3. Global trends to support freshwater conservation

Mapping species distributions is considered important for conservation efforts because it increases understanding of the spatial patterns of endemism and vulnerability. Species mapping may be conducted along with an inventory of current and future landscape-scale anthropogenic stressors. Understanding the global extent of freshwater habitat alteration is important to prioritize areas for protection and restoration while finding global development pathways that balance human demands (e.g., dam construction) with freshwater ecosystem needs [109]; however, a key challenge to mapping freshwater habitat alteration is lack of

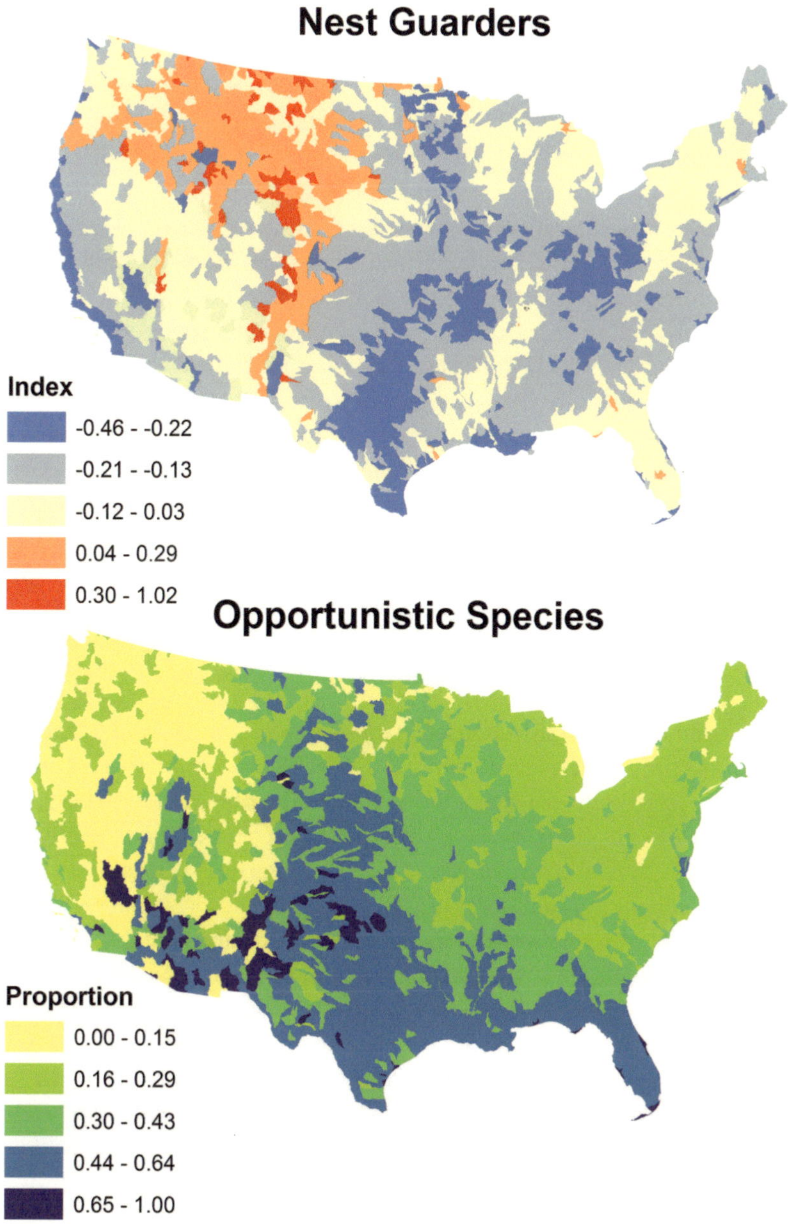

Figure 12. Two examples of species trait biogeography maps for US fish species. Pools of species within watersheds are summarized by their trait values, e.g. averages (nest guarder index) or by proportions of species possessing a trait or having a life history strategy (proportion of opportunistic species). Data from [116].

understanding about how anthropogenic activities propagate impacts in freshwater environments. Freshwaters are influenced by upstream drainage networks, the surrounding landscape, and hence, are recipients of upstream land activities, all of which creates a challenge in modeling, mapping, and understanding conservation challenges [6].

Recently, much progress has been made in understanding the extent and current state of global freshwater habitat alteration due to dam construction and extractive uses of water. Flow regulation and fragmentation were first examined for global large river systems by assessing the percentage of annual runoff captured by reservoirs and the longest mileage of rivers running unobstructed within each basin [110]. The authors found that over half of all large basins in the world are affected by dam fragmentation and/or regulation. Subsequently, reference [111] examined global river flow alterations by using a global water model, WaterGAP, to simulate the effects of reservoirs and withdrawals on river discharges at the 0.5° cell resolution. These were important studies, but properly assessing global impacts of dams and reservoirs required spatially explicit analysis in river networks, which entailed better representation of reservoirs in relation to hydrographic features [11]. The latest estimate suggests that 575,900 river kilometers or 7.6% of the world's rivers have flows regulated by reservoirs [11]. All the above studies provided relatively simplistic indicators of impacts from dams on river environments, which may not translate into predictions of potential biodiversity impacts [109]. In response, Grill et al. [109] developed novel indicators, a river fragmentation index and river regulation index, to examine holistic impacts of dams on major basins of the world currently and planned in the future. Grill et al. [109] concluded that 48% of global river volume is severely impacted by reservoirs and that number would increase to 93% if all dams planned and under construction are completed. Other approaches to quantify widespread anthropogenic alterations to aquatic landscapes also includes historical spatial inventories of waterbodies and habitat loss (e.g., [112])

Examining observed or potential responses of species to environmental change through the lens of species traits provides a mechanism to link species conservation needs to habitat alteration [113, 114]. Species traits are characteristics that describe the life history, ecology, and behavior of organisms. As the name suggests, the field of trait biogeography links species trait values with their spatial distributions [115, 116] (**Figure 12**). This provides a powerful tool to assess or predict individual, community, or regional species pool responses to habitat alterations. For example, by synthesizing global dam occurrences and fish traits in freshwater ecoregions, several fish taxa that were at high risk of species loss could be identified [69]. Several databases are available that provide rich information on species traits. For example, FishBase provides information on taxonomy, conservation status, biology, trophic ecology, and life history for >33,000 freshwater and marine fish species [67]. For North America, the Fish Traits database provides life history information, trophic attributes, reproductive ecology, habitat associations, and salinity/temperature information for >800 native and exotic freshwater fish species [113].

4. Conclusions and implications for biodiversity conservation

Recent developments in global freshwater habitat and biodiversity mapping products (and the rate at which they are updated) is encouraging for future conservation efforts. Assessing the conservation status of species and prioritizing areas of the globe for protection will continue to rely on spatially comprehensive and contiguous inventories of habitats, the biota they support, and evaluation of the degree of alteration at progressively higher spatial resolutions.

http://dx.doi.org/10.5772/intechopen.70296

Metrics are needed that translate anthropogenic stressors into meaningful measures of global habitat alterations in to freshwater systems. Depicting these relationships is challenging for freshwater ecosystems because they are inherently tied to upstream landscape processes. In turn, the field of trait biogeography shows promise in providing a predictive template to convert habitat alterations into specific biodiversity concerns.

While many nations have their own freshwater mapping initiatives conducted at relatively high resolutions (e.g., the US's NHD and NatureServe projects), many underdeveloped nations experiencing intense pressures from development (e.g., Brazil) are likely to rely on external globally-derived products to inform conservation efforts. Even so, local conservation efforts require more spatial fidelity to guide future development pathways. In particular, the Amazon basin is experiencing rapid hydropower development without proper knowledge of the full diversity and geography of fish, invertebrates, and amphibians, or the strategies needed to prevent extinction of these organisms during energy expansion [72]. The development and justification of global reserves for biodiversity conservation will also be contingent upon the accuracy and resolution of aquatic habitats and the organisms they support. New advances in our observation of earth (e.g. through remote sensing), provide opportunities for filling some of these gaps; however, understanding global biodiversity patterns at high resolutions will require exploring local knowledge bases and building predictive models before they disappear.

Acknowledgements

We thank the countless individuals who have recorded observations of species for hundreds of years and those who make spatial data available to support applications in biogeography. Special thanks to Esther Parish for providing comments on earlier versions of this manuscript. This work was funded by the U.S. Department of Energy, Office of Energy Efficiency & Renewable Energy, Water Power Technologies Office and by program development funds provided by the Climate Change Science Institute at Oak Ridge National Laboratory.

Author details

Ryan A. McManamay*, Natalie A. Griffiths, Christoper R. DeRolph and Brenda M. Pracheil

*Address all correspondence to: mcmanamayra@ornl.gov

Environmental Sciences Division, Oak Ridge National Laboratory, Oak Ridge, TN, USA

References

[1] Edie SM, Smits PD, Jablonski D. Probabilistic models of species discovery and biodiversity comparisons. PNAS. 2017;**114**:3666-3671

[2] Mora C, Tittensor DP, Adl S, Simpson AGB, Worm B. How many species are there on Earth and in the Ocean? PLoS Biology. 2011;**9**(8):e1001127. DOI: 10.1371/journal.pbio.1001127

[3] Wake DB, Vredenburg VT. Are we in the midst of the sixth mass extinction? A view from the world of amphibians. PNAS. 2008;**105**:11466-11473

[4] Barnosky AD, Matzke N, Tomiya S, Wogan GOU, Swartz B, et al. Has the Earth's sixth mass extinction already arrived? Nature. 2011;**471**:51-57. DOI: 10.1038/nature09678

[5] IUCN. The IUCN Red List of Threatened Species. Summary Statistics [Internet]. Available from: http://www.iucnredlist.org/about/summary-statistics [Accessed: 11-04-2017]

[6] Dudgeon D, Arthington AH, Gressner MO, Kawabata Z-I, Knowler DJ, et al. Freshwater biodiversity: Importance, threats, status and conservation challenges. Biological Reviews. 2006;**81**:163-182. DOI: 10.1017/S1464793105006950

[7] Balain EV, Segers H, Lévèque C, Martens K. The Freshwater Animal Diversity Assessment: An overview of the results. Hydrobiologia. 2008;**595**:627-637

[8] Olson, DM, Dinerstein E, Wikramanayake ED, Burgess N, Powell N, et al. Terrestrial ecoregions of the world: A new map of life on Earth. Bioscience. 2001;**51**:933-938

[9] Lehner B, Verdin K, Jarvis A. New global hydrography derived from spaceborne elevation data. Eos Transactions of the American Geophysical Union. 2008;**89**:93-94

[10] Abell R, et al. Freshwater ecoregions of the world: A new map of biogeographic units for freshwater biodiversity conservation. BioScience. 2008;**58**:403-414

[11] Lehner B, Reidy Liermann C, Revenga C, Vörösmarty C, Fekete B, et al. High-resolution mapping of the world's reservoirs and dams for sustainable river-flow management. Frontiers in Ecology and the Environment. 2011;**9**:494-502. DOI: 10.1890/100125

[12] Lehner B, Grill G. Global river hydrography and network routing: Baseline data and new approaches to study the world's large river systems. Hydrological Processes. 2013;**27**:2171-2186

[13] Verpoorter C, Kutser T, Seekell DA, Tranvik LJ. A global inventory of lakes based on high-resolution satellite imagery. Geophysical Research Letters. 2014;**41**:6396-6402

[14] Lehner B, Döll P. Development and validation of a global database of lakes, reservoirs and wetlands. Journal of Hydrology. 2014;**296**:L1-L22

[15] Cole JJ, Prarie YT, Craco NF, WH McDowell, et al. Plumbing the global carbon cycle: Integrating inland waters into the terrestrial carbon budget. Ecosystems. 2007;**10**:171-184

[16] Butman D, Raymond PA. Significant efflux of carbon dioxide from streams and rivers in the United States. Nature Geoscience. 2011;**4**:839-842

[17] Raymond PA, et al. Global carbon dioxide emissions from inland waters. Nature. 2013;**503**:355-359

[18] Collen B, Whitton F, Dyer EE, Baillie JEM, Cumberlidge N. Global patterns of freshwater species diversity, threat and endemism. Global Ecology and Biogeography. 2014;**23**:40-51

[19] Richman NI, Bohm M, Adams SB, Alvarez F, Bergey EA. Multiple drivers of decline in the global status of freshwater crayfish (Decapoda: Astacidae). Philosophical Transactions of the Royal Society B. 2015;**370**:2014

[20] Downing JA, Cole JJ, Duarte CM, Middelburg JJ, et al. Global abundance and size distribution of streams and rivers. Inland Waters. 2012;**2012**:229-236

[21] Lehner B, Döll P. Development and validation of a global database of lakes, reservoirs and wetlands. Journal of Hydrology. 2004;**296**:1-22

[22] Downing JA, Prairie YT, Cole JJ, Duarte CM, Tranvik LJ, Striegl, RG, McDowell WH, Kortelainen P, Caraco NF, Melack JM, et al. The global abundance and size distribution of lakes, ponds, and impoundments. Limnology and Oceanography. 2006;**51**:2388-2397

[23] Dodds PS, Rothman DH. Geometry of river networks. III. Characterization of component connectivity. Physical Review. 2000;**63**:016117

[24] Reis AH. Constructal views of scaling laws of river basins. Geomorphology. 2006; **78**:201-206

[25] Leopold LB. Rivers. American Scientist. 1962;**50**(4):511-537

[26] U.S. Defense Mapping Agency. Development of the Digital Chart of the World. Washington, D.C., USA: Government Printing Office; 1992

[27] American Association for the Advancement of Science. Digital Chart of the World [Internet]. 2013. Available from: https://www.aaas.org/page/digital-chart-world [Accessed: 12-04-2017]

[28] US Geological Survey. Long Term Archive. HYDRO1K [Internet]. 2015. Available from: https://lta.cr.usgs.gov/HYDRO1K [Accessed: 12-04-2017]

[29] Aufdenkampe AK, Mayorga E, Raymond PA, Melack JM, Doney SC et al. Riverine coupling of biogeochemical cycles between land, oceans, and atmosphere. Frontiers in Ecology and the Environment. 2011;**9**:53-60

[30] Downing JA. Plenary Lecture. Global limnology: Up-scaling aquatic services and process to planet Earth. Verhandlungen des Internationalen Verein Limnologie. 2009;**30**: 1149-1166

[31] Wu H, Kimball JS, Mantua N, Stanford J. Automated upscaling of river networks for macroscale hydrological modeling. Water Resources Research. 2011;**47**:W03517

[32] Wu H, Kimball JS, Li H, Huan M, Leung LR, Adler RF. A new global river network database for macroscale hydrologic modeling. Water Resources Research. 2012;**48**:W09701

[33] Döll P, Lehner B. Validation of a new global 30-min drainage direction map. Journal of Hydrology. 2002;**258**:214-231

[34] Horizon Systems. NHDPlus Version 2. [Internet]. 2017. Available from: http://www.horizon-systems.com/NHDPlus/index.php [Accessed: 12-04-2017]

[35] McKay L, Bondelid T, Dewald T, Johnston J, Moore R, Rea A. NHDPlus Version 2: User Guide. Horizon Systems. 2012. p. 180. Available from: ftp://ftp.horizon-systems.com/NHDplus/NHDPlusV21/Documentation/NHDPlusV2_User_Guide.pdf [Accessed: 12-04-2017]

[36] US Geological Survey. Hydrography. National Hydrography Dataset. Watershed Boundary Dataset [Internet]. 2017. Available from: https://nhd.usgs.gov/ [Accessed: 12-04-2017]

[37] Kelly CA, Rudd JWN, St. Louis V, Moore T. Turning attention to reservoir surfaces, a neglected area in greenhouse studies. EOS, Transactions of the American Geophysical Union. 1994;**75**:332-333

[38] Pearce F. Trouble bubbles for hydropower. New Science. 1996;**150**:28-31

[39] Meybeck M. Global distribution of lakes. In: Lerman A, Imboden DM, Gat JR, editors. Physics and Chemistry of Lakes. Berlin: Springer; 1995. pp. 1-36

[40] McDonald CP, Rover JA, Stets EG, Striegl RG. The regional abundance and size distribution of lakes and reservoirs in the United States and implications for estimates of global lake extent. Limnology and Oceanography. 2012;**57**:597-606

[41] St. Louis VL, Kelly CA, Ducemin E, Rudd JWM, Rosenberg DM. Reservoir surfaces as sources of greenhouse gases to the atmosphere: A global estimate. BioScience. 2000; **50**:766-775

[42] Messager ML, Lehner B, Grill G, Nedeva I, Schmitt O. Estimating the volume and age of water stored in global lakes using a geo-statistical approach. Nature Communications. 2016;**7**:13603. DOI: 10.1038/ncomms13603

[43] Travnik LJ, et al. Lakes and reservoirs as regulators of carbon cycling and climate. Limnology and Oceanography. 2009;**54**:2298-2314

[44] Seekell DA, Pace ML. Does the Pareto distribution adequately describe the size-distribution of lakes. Limnology and Oceanography. 2011;**56**:350-356

[45] Cowardin LM, Carter V, Golet FC, LaRoe ET. Classification of wetlands and deepwater habitats of the United States. FWS/OBS-79/31. Washington, DC: U.S. Fish & Wildlife Service Pub; 1979

[46] Navid D. The international law of migratory species: The Ramsar convention. Natural Resources Journal. 1989;**29**:1001-1016

[47] Dugan, P. Wetlands in Danger: A World Conservation Atlas. UK: Oxford University Press; 1993

[48] US Geological Survey. Long Term Archive. Global Land Cover Characteristics Data Base Version 2.0 [Internet]. 2015. Available from: https://lta.cr.usgs.gov/glcc/globdoc2_0 [Accessed: 12-04-2017]

[49] Land Process Distributed Archive Center. MODIS Overview [Internet]. 2014. Available from: https://lpdaac.usgs.gov/dataset_discovery/modis [Accessed: 12-04-2017]

[50] Matthews E, Fung I. Methane emission from natural wetlands: Global distribution, area, and environmental characteristics of sources. Global Biogeochemical Cycles. 1987;**1**:61-86

[51] Cogley JG. GGHYDRO: Global Hydrographic Data, Release 2.1. [Dataset]. Peterborough, Ont., Canada: Department of Geography, Trent University; 1994. Available from: http://people.trentu.ca/~gcogley/glaciology/glglgghy.htm [Accessed: 12-04-2017]

[52] Stillwell-Soller LM, Klinger LF, Pollard D, Thompston SL. The Global Distribution of Freshwater Wetlands. Boulder, CO: NCAR Technical Note TN-416+STR, National Center for Atmospheric Research; 1995

[53] Rebelo LM, Finlayson CM, Nagabhatla N. Remote sensing and GIS for wetland inventory, mapping and change analysis. Journal of Environmental Management. 2009;**90**:2144-2153

[54] Ramsar Convention Secretariat. Wetland Inventory: A Ramsar Framework for Wetland Inventory and Ecological Character Description. Ramsar Handbooks for the Wise Use of Wetlands. 4th ed. Gland, Switzerland: Ramsar Convention Secretariat; 2010

[55] Williams M. Wetlands: A Threatened Landscape. Oxford, UK: Blackwell; 1991

[56] Mitsch WJ, Gosselink JG. Wetlands. 3rd ed. New York: Wiley; 2000

[57] Finlayson CM, Davidson NC. Global Review of Wetland Resources and Priorities for Wetland Inventory: Summary Report. Australia: Wetlands International, The Netherlands, and the Environmental Research Institute of the Supervising Scientists. 1999. Available online at: http://www.wetlands.org/inventoryand/GRoWI/welcome.html [Accessed 12-04-2017]

[58] Homer CG, Dewitz JA, Yang L, Jin S, Danielson P, et al. Completion of the 2011 National Land Cover Database for the conterminous United States-Representing a decade of land cover change information. Photogrammetric Engineering and Remote Sensing. 2015;**81**:345-354

[59] US Fish, Wildlife Service. National Wetlands Inventory [Internet]. 2017. Available from: https://www.fws.gov/wetlands/ [Accessed 12-04-2017]

[60] Dahl TE. Status and Trends of Wetlands in the Conterminous United States 2004 to 2009. Washington, D.C.: U.S. Department of the Interior; Fish and Wildlife Service; 2011. p. 108

[61] Turner W, Spector S, Gardiner N, Fladeland M, Sterling E, Steininger M. Remote sensing for biodiversity science and conservation. Trends in Ecology and Evolution. 2003;**18**:306-314. DOI: 10.1016/S0169-5347(03)00070-3

[62] US Department of Interior. Population Surveys of Nesting American White Pelicans [Internet]. 2015. Available from: https://eros.usgs.gov/doi-remote-sensing-activities/2015/population-surveys-nesting-american-white-pelicans [Accessed: 12-04-2017]

[63] Oberdorff T, Gue´gan JF, Hugueny B. Global scale patterns of fish species richness in rivers. Ecography. 1995;**18**:345-352

[64] Amarasinghe US, Welcomme RL. An analysis of fish species richness in natural lakes. Environmental Biology of Fishes. 2002;**65**:327-339

[65] International Lake Environment Committee (ILEC). World Lake Database [Internet]. 2017. Available from: http://wldb.ilec.or.jp/ [Accessed 12-04-2017]

[66] Xenopoulous MA, Lodge DM, Alcamo J, Märker M, Shulze K. Scenarios of freshwater fish extinctions from climate change and water withdrawal. Global Change Biology. 2005;**11**:1557-1564. DOI: 10.1111/j.1365-2486.2005.01008.x

[67] Froese R, Pauly D. FishBase [Internet]. 2017. Available from: http:/www.fishbase.org [Accessed: 2017-04-2012]

[68] Oberdorff T, Tedesco PA, Hugueny B, Leprieur F, Beauchard O, et al. Global and regional patterns in riverine fish species richness: A review. International Journal of Ecology. 2011;**2011**:967631. DOI: 10.1155/2011/967631

[69] Liermann CR, Nilsson C, Robertson J, Ng RY. Implications of dam obstruction for Global freshwater fish diversity. BioScience. 2012;**62**:539-548

[70] Brosse S, Fish-SPRICH: A database of freshwater fish species richness throughout the World. Hydrobiologia. 2013;**700**:343-349

[71] Toussaint A, Charpin N, Brosse S, Villéger S. Global functional diversity of freshwater fish is concentrated in the Neotropics while functional vulnerability is widespread. Nature Scientific Reports. 2016;**6**:22125. DOI: 10.1038/srep22125

[72] Winemiller KO, McIntyre P, Castello L, Fluet-Chouinard E, Giarrizzo T, et al. Balancing hydropower and biodiversity in the Amazon, Congo, and Mekong: Basin-scale planning is needed to minimize impacts in mega-diverse rivers. Science. 2016;**351**:128-129

[73] International Union for the Conservation of Nature. Spatial Data Download. Red List Spatial Data [Internet]. 2017. Available from: http://www.iucnredlist.org/technical-documents/spatial-data [Accessed 12-04-2017]

[74] Stuart SN, Chanson JS, Cox NA, Young BE, Rodrigues SL, et al. Status and trends of amphibian declines and extinctions worldwide. Science. 2004;**306**:1783-1786

[75] International Union for the Conservation of Nature. Amphibian Assessment Process [Internet]. 2016. Available from: http://www.iucnredlist.org/initiatives/amphibians/process [Accessed 12-04-2017]

[76] Gallant AL, Klaver RW, Casper GS, Lannoo MJ. Global rates of habitat loss and implications for amphibian conservation. Copeia. 2007;**2007**:967-979

[77] Sodhi NS, Bickford D, Diesmos AC, Lee TM, Koh LP, et al. Measuring the meltdown: Drivers of Global amphibian extinction and decline. PLoS One. 2008;**3**:e1636. DOI: 10.1371/journal.pone.0001636

[78] Rödder D, Kielgast J, Bielby J, Schmidtlein S, Bosch J, et al. Global amphibian extinction risk assessment for the Panzootic Chytrid fungus. Diversity. 2009;**1**:52-66. DOI: 10.3390/d1010052

[79] Hof C, Araújo MB, Jetz W, Rahbek C. Additive threats from pathogens, climate and land-use change for global amphibian diversity. Nature. 2011;**480**:516-520

[80] Ficetola GF, Rondinini C, Bonardi A, Katariya V, Padoa-Schioppa E, Angulo A. An evaluation of the robustness of global amphibian range maps. Journal of Biogeography. 2014;**41**:211-221

[81] Global Biodiversity Information Facility. Free and Open Access Biodiversity Data [Internet]. 2017. Available from: http://www.gbif.org/ [Accessed 12-04-2017]

[82] Check List. Journal of Species Lists and Distribution [Internet]. 2017. Available from: http://www.checklist.org.br/about [Accessed 12-04-2017]

[83] Graf DL, Cummings KS. Review of the systematics and global diversity of freshwater mussel species (Bivalvia: Unionoida). Journal of Molluscan Studies. 2007;**73**:291-314. DOI: 10.1093/mollus/eym029

[84] Mussel Project. The Freshwater Mussels (Unionoida) of the World (and other less consequential bivalves) [Internet]. 2017. Available from: http://mussel-project.uwsp.edu/index.html [Accessed 12-04-2017]

[85] Nobles T, Zhang Y. Biodiversity loss in freshwater mussels: Importance, threats, and solutions. In: Biodiversity loss in a Changing Planet. InTechOpen; Rijeka, Croatia. 2011. pp. 137-162. DOI: 10.5772/25102. ISBN:978-953-307-707-9

[86] Crandall KA, Buhay JE. Global diversity of crayfish (Astacidae, Cambaridae, and Parastacidae—Decapoda) in freshwater. Hydrobiologia. 2008;**595**:295-301

[87] Rodrigues ASL, Andelman SJ, Bakarr MI, Boitani L, Brooks TM, et al. Effectiveness of the global protected area network in representing species diversity. Nature. 2004;**428**:640-643

[88] Rodrigues ASL, Akcakaya HR, Andelman SJ, Bakarr MI, Boitani L, et al. Global gap analysis: Priority regions for expanding the Global Protected-Area Network. BioScience. 2004;**54**:1092-1100

[89] Grenyer R, Orme CDL, Jackson SF, Thomas GH, Davies RG, et al. Global distribution and conservation of rare and threatened vertebrates. Nature. 2006;**444**:93-96. DOI: 10.1038/nature05237

[90] McGeoch MA, Butchart SHM, Spear D, Marais E, Kleynhans EJ. Global indicators of biological invasion: Species numbers, biodiversity impact and policy responses. Diversity and Distributions. 2010;**16**:95-108

[91] Pim SL, Jenkins CN, Abell R, Brooks TM, Gittleman JL, et al. The biodiversity of species and their rates of extinction, distribution, and protection. Science. 2014;**344**:1246752-1-1246752-10

[92] Jenkins CN, Van Houtan KS, Pimm SL, Sexton JO. US protected lands mismatch biodiversity priorities. Proceedings of the National Academy of Sciences. 2015;**112**:5081-5086

[93] NatureServe. Data, Maps, and Tools [Internet]. 2017. Available from: http://www.natureserve.org/conservation-tools/data-maps-tools [Accessed 12-04-2017]

[94] BirdLife International [Internet]. 2017. Available from: http://www.birdlife.org/ [Accessed 12-04-2017]

[95] US Geological Survey. Digital representations of tree species range maps from "Atlas of United States Trees" by Elbert L. Little, Jr. (and other publications) [Internet]. 2015. Available from: https://esp.cr.usgs.gov/data/little/ [Accessed 12-04-2017]

[96] iSpot. Communities-Global [Internet]. 2017. Available from: http://www.ispotnature.org/communities/global [Accessed 12-04-2017]

[97] Life Mapper. The Lifemapper Project [Internet]. 2017. Available from: http://lifemapper.org [Accessed 12-04-2017]

[98] Ready J, Kaschner AB, South PD, Eastwood T, Rees J, et al. Predicting the distributions of marine organisms at the global scale. Ecological Modelling. 2010;**221**:467-478. DOI: 10.1016/j.ecolmodel.2009.10.025

[99] AquaMaps. Standardized distribution maps for over 25,000 species of fishes, marine mammals and invertebrates [Internet]. 2016. Available from: http://www.aquamaps.org/ [Accessed 12-04-2017]

[100] Phillips SJ, Anderson RP, Schapire RE. Maximum entropy modeling of species geographic distributions. Ecological Modelling. 2006;**190**:231-259

[101] Domisch S, Amatulli G, Jetz W. Near-global freshwater-specific environmental variables for biodiversity analyses in 1 km resolution. Scientific Data. 2015;**2**:150073. DOI: 10.1038/sdata.2015.73

[102] Dauwalter DC, Rahel F. Distribution modelling to guide stream fish conservation: An example using the mountain sucker in the Black Hills National Forest, USA. Aquatic Conservation: Marine and Freshwater Ecosystems. 2008;**18**:1263-1276. DOI: 10.1002/aqc.940

[103] Steen PJ, Zorn TG, Seelbach PW, Schaeffer JS. Classification tree models for predicting distributions of Michigan stream fish from landscape variables. Transactions of the American Fisheries Society. 2008;**37**:976-996. DOI: 10.1577/T07-119.1

[104] Worthington TA, Brewer SK, Grabowski TB, Mueller J. Backcasting the decline of a vulnerable Great Plains reproductive ecotype: Identifying threats and conservation priorities. Global Change Biology. 2014;**20**:89-102. DOI: 10.1111/gcb.12329

[105] Huang J, Frimpong EA. Limited transferability of stream-fish distribution models among river catchments: Reasons and implications. Freshwater Biology. 2016;**61**:729-744. DOI: 10.1111/fwb.12743

[106] Huang J, Frimpong EA. Using historical atlas data to develop high-resolution distribution models of freshwater fishes. PLoS One. 2015;**10**:e0129995. DOI: 10.1371/journal.pone.0129995

[107] Cao Y, Hinz L, Metzke B, Stein J, Holtrop A. Modeling and mapping fish abundance across wadeable streams of Illinois, USA, based on landscape-level environmental variables. Canadian Journal of Fisheries and Aquatic Sciences. 2016;**73**:1031-1046. DOI: 10.1139/cjfas-2015-0343

[108] DeRolph CR, Nelson SA, Kwak TJ, Hain EF. Predicting fine-scale distributions of peripheral aquatic species in headwater streams. Ecology and Evolution. 2015;**5**:152-163. DOI: 10.1002/ece3.1331

[109] Grill G, Lehner B, Lumsdon AE, MacDonald GK, Zarfl C, Lierman CR. An index-based framework for assessing patterns and trends in river fragmentation and flow regulation by global dams at multiple scales. Environmental Research Letters. 2015;**10**:015001. DOI: 10.1088/1748-9326/10/1/015001

[110] Nilsson C, Reidy CA, Dynesius M, Revenga C. Fragmentation and flow regulation of the World's large river systems. Science. 2005;**308**:405-408. DOI: 10.1126/science.1107887

[111] Döll P, Fiedler K, Zhang J. Global-scale analysis of river flow alterations due to water withdrawals and reservoirs. Hydrologic Erath System Sciences. 2009;**13**:2413-2432. www.hydrol-earth-syst-sci.net/13/2413/2009/

[112] Im RY, Kim JY, Joo GJ, Do Y. Process of wetland loss in the lower Nakdong River, South Korea. Applied Ecology and Environmental Research. 2017;**15**:69-78

[113] Frimpong EA, Angermeier. FishTraits: A database of ecological and life-history traits of freshwater fishes of the United States. Fisheries. 2009;**34**:487-495

[114] Frimpong EA, Angermeier PL. Trait based approaches in the analysis of stream fish communities. In: Gido KB, Jackson DA, editors. Community Ecology of Stream Fishes: Concepts, Approaches, and Techniques. Bethesda, Maryland, USA: American Fisheries Society, Symposium 73; 2010. pp. 109-136

[115] Mims MC, Olden JD, Shattuck ZR, Poff NL. Life history trait diversity of native freshwater fishes in North America. Ecology of Freshwater Fish. 2010;**19**:390-400

[116] McManamay RA, Frimpong EA. Hydrologic filtering of fish life history strategies across the United States: Implications for stream flow alteration. Ecological Applications. 2015;**25**:243-263

[117] Convention on Biological Diversity [Internet]. 2017. Available from: https://www.cbd.int/. [Accessed 12-04-2017]

Section 3

Applied Biogeography and Regional Patterns

Chapter 5

Plant Antiherbivore Defense in Diverse Environments

Alina Morquecho-Contreras,
Carmen Zepeda-Gómez and
Hermilo Sánchez-Sánchez

Additional information is available at the end of the chapter

http://dx.doi.org/10.5772/intechopen.70418

Abstract

Herbivores can damage plant productivity and fitness; plants have improved defensive traits, such as chemical defenses. Plant species produce specific defensive traits in response of diverse risk factor generated by herbivores. In this chapter, we analyze and compare the defensive traits used by plants in different habitats: aquatic ecosystems, temperate forest, and rainforest. In aquatic environments, the number of herbivores is scarce, and plants develop biomass and restrict defensive compound production. At the terrestrial environment, plants need to accumulate defensive traits for an eventual attack. But the number and quantity of those traits depend on biotic and abiotic factors. In temperate forest, plants have a low growth, and herbivore diversity is low, because there are a few number of defensive traits but in great quantity to guarantee plant survival. In contrast, at tropical forest there is a great herbivore diversity, and plants have a quick growth; thus they develop a great variety of defensive traits. There are substantial differences in plant defensive strategies at different environments. Usually, the aquatic plants use water-soluble and diffusible compounds; plants in rainforest use a plethora of chemical defenses, and in temperate forest, plants utilize physical barriers, resins, and terpenes.

Keywords: aquatic environment, terrestrial environment, temperate forest, tropical forest, plant defenses

1. Introduction

Photosynthetic organisms are the primary producers; they are essential for correct function of all ecosystems; nevertheless, these organisms are susceptible to be attacked by different herbivores, and they can perform various defensive measures: allocate resources to protect themselves from microbes, competitors, ultraviolet (UV) rays, and predators [1], reduce plant tissue quality, and produce chemical and mechanical defenses [2, 3].

Throughout 350 million years, plants and insects have been keeping a close relationship [4] resulting in an efficient defense system in plants that can recognize signals from herbivore and activate the plant immune response against them. To arrest herbivore attack, plants produce specialized metabolites with negative physiological effects against herbivores, such as toxins, deterrent, dissuasive, and/or no nutrition [5].

Metabolites that implicate in defense against herbivores can be modified by biotic and abiotic factors, such as humidity, altitudinal gradient, nutrient availability, herbivores diversity, etc., [6]. Then, we asked: at distinct environments, are plant defense mechanisms the same? If they are different, are there some recognizable patterns at separate environments?

To answer, we select three very distinct ecosystems to compare plant defense traits: the first great difference is between aquatic and terrestrial environments, and the latter we divide in tempered and tropical forest. The objective is recognizing the ecological and evolutionary diversification of plant defense traits at distinct environments.

2. Aquatic environments

In aquatic environments, there is a great diversity of photosynthetic organisms that interact and maintain complex ecological relationships with herbivores. In aquatic habitats, these interactions are considered very important since they affect the nutrient cycle and energy flows of food chains [7, 8].

Generally, when a plant is attacked, its defense mechanisms are activated through the production of diverse compounds generically termed plant secondary metabolites (PSM) [9]. It has been recognized that the secondary compounds may either serve as feeding deterrents or attractants in terrestrial plant-animal interactions or function as allelopathic chemicals or antibiotics; the same evolutionary pressures responsible for the many biologically active compounds found in terrestrial vegetation have been predicted to have parallels in marine [1, 10] and freshwater vegetation [11]. However, it is possible that the constraints in aquatic habitat lead to some differences in the production and action of these natural compounds.

2.1. Marine environments

The primary producers most widely distributed in marine habitats are the seaweeds (red, brown, and green algae); these photosynthetic organisms have developed several defenses in response to herbivores, for example, by having a resistant or unpalatable physical structure or a morphology that makes the feed difficult for the herbivore or by having spatial and temporally diverse stages of life cycle and by the production of chemical defense against herbivores ranging from unpalatable to toxic. Marine algae are known to produce a wide range of secondary metabolites with various biological actions [1], many of them with medicine and agriculture human uses [12].

On the sea, the most common grazers are generalist such as fishes, sea urchins, gastropods, polychaetes, and a great variety of marine crustaceans [7], which usually feed on the stalks of seaweed. Over 2400 natural products have been isolated from marine red, brown, and green algae, the majority are terpenoids and acetogenins; very few nitrogenous compounds have been isolated. In general, these compounds occur in relatively low concentration (0.2–2.0% dry mass), even so several ecological roles have been documented, and some of them are produced as protection against grazing [13, 14].

Brown algae produce about 1000 secondary metabolites, terpenoids and acetogenins are the most frequent, and they are the only seaweeds that produce polyphenolic compounds [12]. Polyphenolic compounds may function like terrestrial tannins, but they are structurally different so they are often termed phlorotannins to distinguish from them [12, 15, 16]. Phlorotannins are usually associated with a chemical defense: protection against grazing, pathogen attack, epiphytism, microfouling, and ultraviolet (UV) damages [16]. In red algae, the greatest variety of secondary metabolites is found, about 1240 reported; in Rhodophyta, all classes of compounds except phlorotannins can be found; most of them are halogenated (methanes, haloketones, phenolics, and complex terpenes) [15] recognized as antibacterial, antifungal, antiviral, anti-inflammatory, antiproliferative, antifouling, antifeedant, cytotoxic, ichthyotoxic, and insecticidal activity [17]. In contrast, the green algae are the ones with the least secondary metabolites isolated; about 290 are known and most of them are sesquiterpenes and diterpenes; only few species produced halogenated compounds [18].

These diverse compounds are consumed directly by the herbivore when it feeds algae, but many of them may be released into the aquatic environment during algal growth or at cell lysis. In the last cases, chemical information is transmitted by diffusion and adventive lamina flow [19]; a major problem in the aquatic environment is dilution of the secreted products, so small molecules are favored because of their faster diffusion.

As we see, the natural compound production differs among seaweeds, as well as between and within species [7, 20], these differences suggest separate historical origin [11]. In the same way, the type and quantity of secondary compounds in algae differ from vascular plants; in algae, the absence of alkaloids and the presence of halogens compounds have been detected, contrary to terrestrial plants [7].

In marine environments, sea grasses are the only true submerged angiosperm, and as vascular plants, they are more complex morphologically and physiologically than algae; they produce some secondary compounds against herbivores, such as phenolic acids, phenolic acid sulfate esters, and sulfated flavonoids [21].

2.2. Freshwater and continental environments

In continental and freshwater environments, angiosperms are more abundant than macroalgae; therefore, they contribute significantly to primary productivity, and they maintain numerous interactions with aquatic consumers such as birds, mammals, fishes, crayfish, insects, and mollusks [22, 23]. For a long time, it was considered that the herbivory on freshwater macrophytes was infrequent and with minimal impact [24, 25]. Contrary to this point of view, a growing body of evidence suggests that the evolutionary and ecological importance of herbivory occurs in an aquatic context as in terrestrial habitats [9, 11]. Interactions between herbivores and aquatic plants have been reported in a wide range of habitat types, including freshwater lakes, rivers, estuaries, wetlands, and shallow seas [26, 27]. Accordingly, interactions between herbivores and aquatic plants have global distribution, and herbivores are present wherever submerged, floating, or emergent plants are present [27]. It is a fact that aquatic herbivores have a strong impact on aquatic plant biomass, productivity, and species composition [22, 28]; thus, like in terrestrial angiosperms, selection may favor aquatic plants that have chemical and other types of antiherbivore defenses [9].

Defense and resistance mechanisms against herbivores have been poorly understood in freshwater; even so we now know that freshwater plants are frequently chemically or structurally defended from consumers [29–31]. Structural defenses are more commonly found among upland plants than wetland plants [22]; in some cases, we can find thorns or tough leaves [32]. Chemical defenses are more widespread in macrophytes [23, 31] as well as in various algae, cyanobacteria [22].

Diverse groups of chemical compounds are known in aquatic plants, including alkaloids [33, 34], flavonoids, steroids, saponins, phenolics (including tannins), cyanogenic glycosides, glucosinolates [23, 29], quinines, and essential oils [32]. The different types of chemical defenses can vary between species, localities, time, and environmental conditions [31]. Many of them have not been identified; some studies have found multiple dissuasive components in the chemical extracts analyzed, but the low concentrations or their unstable state makes their identification difficult and therefore their correlation with the dynamics of the aquatic community [35].

In the aquatic environment, plant-herbivore interactions are different from terrestrial ecosystems because water provides different physicochemical conditions compared with air or soil, which should affect the herbivore access and the dispersal of released compounds [36].

2.2.1. Macrophyte growth adaptations

The growth forms of macrophytes are the most significant adaptation to freshwater environments and have important consequences for aquatic plant-herbivore interactions. The structure of the macrophytes and the presence of leaves and flowers above or below the water level determine the access and type of herbivores [36], so structures above the water surface can be consumed by terrestrial herbivores while the submerged parts by aquatic herbivores. Therefore the growth forms may have different mechanisms to prevent herbivory. Compared with terrestrial vegetation, freshwater aquatic plants produce less phenolic compounds, and a different phenolic amount in the aquatic growth forms has been observed. Lodge [22] indicated that the rank of mean phenolic content in wetland plants is tree > floating leaves plants > emergent > submersed > algae. Submerged macrophytes have much lower content than emergent or floating leaved macrophytes [37]. These differences are because emergent plants need more structural tissue, thicker cell walls, and a more complex cuticle to limit evapotranspiration and provide stability; therefore they present structural defenses, while submerged macrophytes are less structurally defended because they have little lignification, thin cuticles to facilitate gas exchange with water, and less exposure to ultraviolet light [36]. As a consequence, interactions with herbivores are modified, fully aquatic leaves of amphibious species and submersed plants exhibited higher grazing rates than aerial leaves, possibly due to a lower structural defense [38].

It is considered that in freshwater plants, constitutive chemical resistance against herbivores are frequent [31, 39, 40] presumably because of a high and lengthy exposure to mostly generalist herbivores [34]. Plants, which would be attacked by generalist herbivores, tend to be defended by a diverse collection of toxins in small concentrations, whereas plants attacked by specialist herbivores tend to employ higher levels of compounds, which reduce digestibility, as it happens in numerous terrestrial plants that are consumed by specialist herbivores.

Aquatic plant-herbivore interactions are highly variable across aquatic ecosystems [11], and we have little information about the presence, levels, types, and function of PSMs; thus we require further analysis in order to make suitable generalizations.

3. Terrestrial ecosystems

The ancestors of terrestrial plants are closely related with charophytes. Plant terrestrialization was preceded by terrestrial algae after aquatic algae, with adaptive mechanisms to live in terrestrial environment such as drought, resistance to UV radiation. In addition, land plants need to increase its body size and cellular differentiation [41].

Another important topic is defensive mechanisms. Aquatic plants have only a few defensive compounds, and their structural defenses are limited because its biomass is constantly renewable. By contrast, terrestrial plants have a plethora of both chemical and structural defensive traits elaborate and accumulate by long-time periods [42]. Nevertheless, at different environment conditions, plants can accumulate diverse molecules or develop distinct structural mechanisms.

Abiotic factors, such as altitude, drought, and nutrient availability, can control defensive traits in plants. At altitudinal gradient, it has been hypothesized that the plant species growing at lower elevations need to invest more in defensive traits because they have greater herbivore pressure, whereas high-elevation plants need less defensive traits [43]. So, as expected at tropical forest, there are more defensive traits than at temperate forest.

Drought slows growth to decrease photosynthetic rate and, in moderate drought, an increase in secondary metabolites is possible, including defensive compounds and structures [44]. In tropical forest, the humidity remains relatively constant, then that factor is not significant.

Plant defense mechanisms are partially due to resources availability; at high resource availability, there are more photosynthesis and growth; but at low resource environments, plant increased defense allocation, because with herbivore attack, it is much more difficult to replace tissue [6, 45, 46].

3.1. Temperate forest

In temperate perennial forest, dominant plants are gymnosperms (see **Figure 1** for distribution). Conifers are dominant vegetation since 200 million years ago, and they appeared 300 million year ago, and during their evolution, they have had few changes. Currently, there are 630 species, which dominate many terrestrial ecosystems, principally in the Northern Hemisphere [47].

Gymnosperms are the major plant lineage with less leaf herbivory, only 0.9%. Low percentage can be explained to tough needle tissue and the presence of terpenoid resins [48]. Herbivory is low in temperate forest tree canopies, and the highest foliar damage occurs in high-quality leaves as youngest [49], it is more probably that the trunk is attacked by bark beetles [3].

At conifer communities, the primary compounds in defense are principally phenolics, terpenoids, and alkaloids, which lay up in the bark [50]. Another line of defense is resin ducts, which confer resistance to insect attack by resin production, flow, and chemical content (**Table 1**) [51].

There are two defense kinds: constitutive, which are expressed all time, even when they are not suffering from damage and induced defenses that enhanced after damage [50, 52]. Among constitutive defenses are resin canals, chemical compounds such as phenolics and therpenes, and the mechanical properties of the cortex that act as a barrier [50, 53]. Those constitutive compounds are nonselective against herbivores, but chemicals produced by induced defense

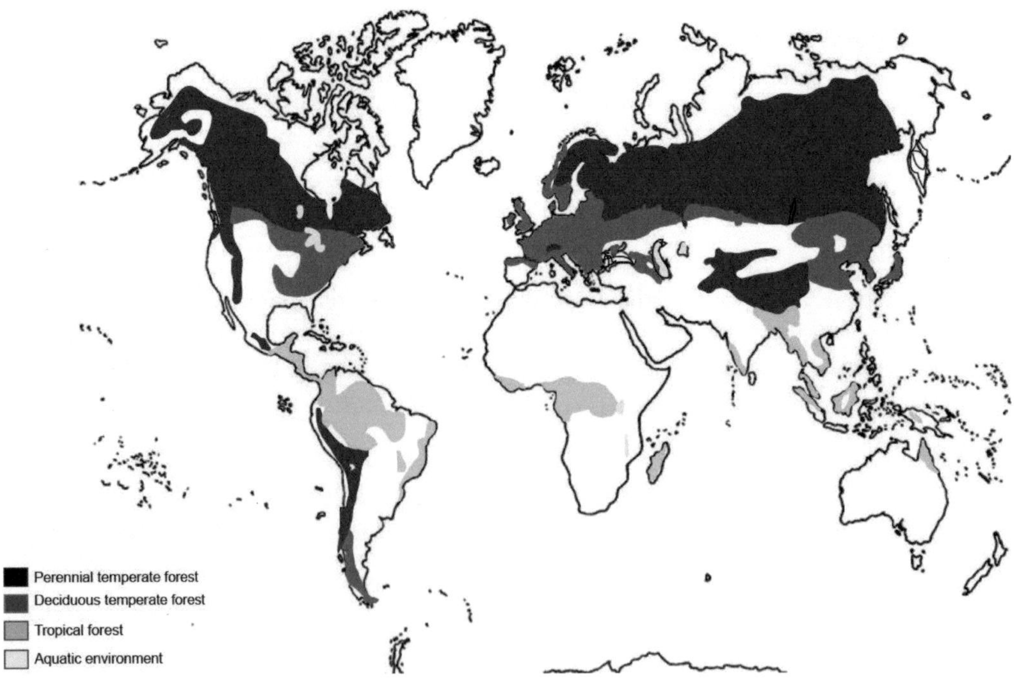

Figure 1. General distribution of different ecosystems in the world.

Environment	Principal plant	Chemical compounds produced
Aquatic environment		
Marine	Red, brown, and green algae	2400 products, the majority terpenoids, acetogenins, phlorotannins (polyphenolics), and halogens
Freshwater	Angiosperms submerged, floating, or emergent	Alkaloids, flavonoids, steroids, saponins, phenolics (including tannins), cyanogenic glycosides, quinines, and essential oils
Temperate forest		
Perennial	Conifers	Terpenoid resins, phenolics, tannins, and alkaloids
Deciduous	Subclass Hamamelidae	Condensed tannins and phenolics
Tropical forest		
	The most diverse plant ecosystem	Great variety of chemical compounds, principally blends, which include phenolics, saponins, amino acids, amides, alkaloids, azoxy glycosides, and terpenes

Table 1. Chemical compounds produced by principal plant types at different environments.

have a wide spectrum and form specific compounds, so they are very diverse and cover a great variety of herbivores [54, 55]. Among induced defenses include phenolic compounds, resin terpenoids, enzymes, PR proteins, and reactive oxygen species [50].

In conifers, resin terpenoids are produced during and after attack of insects, when constitutive duct is active and produces resin, which can flow at great quantity [54, 56].

The bark of conifers has abundant phenolic compounds [57–59]. When phenolics and tannins join to amino acids and proteins, both of them reduce the nutritional value and the ability of insects to digest plant tissues [60]. Constituent phenols can be converted in polyphenolic compounds after herbivore attack, and they are more toxic and specific against herbivores [61].

There are proteins than act as chemical defenses, such as enzymes that degrade components of herbivores such as glucanases and chitinases [62]. Generally, chemical defenses have multiple strategies overlapping that result in a chemical toxic cocktail that stops or destroys an aggressive or virulent attack.

At gymnosperms, multiple overlaying defense systems provide an efficient barrier against a wide range of possible insect attacks. However, conifers remain susceptible to certain organisms that have evolved strategies to overcome the defenses or avoid them. Nevertheless, the remarkable longevity of conifers is a proof to the success of their defense strategies (**Figure 2**) [53].

Another plant community at temperate environment is deciduous forest, particularly the oak forest (distribution in **Figure 1**). In *Quercus robur*, phenolic concentration in leaves increases toward higher elevations with a decrease in leaf damage in comparison with organism at same species that grow in lower altitude [43] that suggest that temperate oak forest is less susceptible to insect damage that tropical forest. In the same way, *Quercus variabilis* total phenolic and total condensed tannin concentration decreases to higher elevation and is more concentrated in juvenile individuals [63].

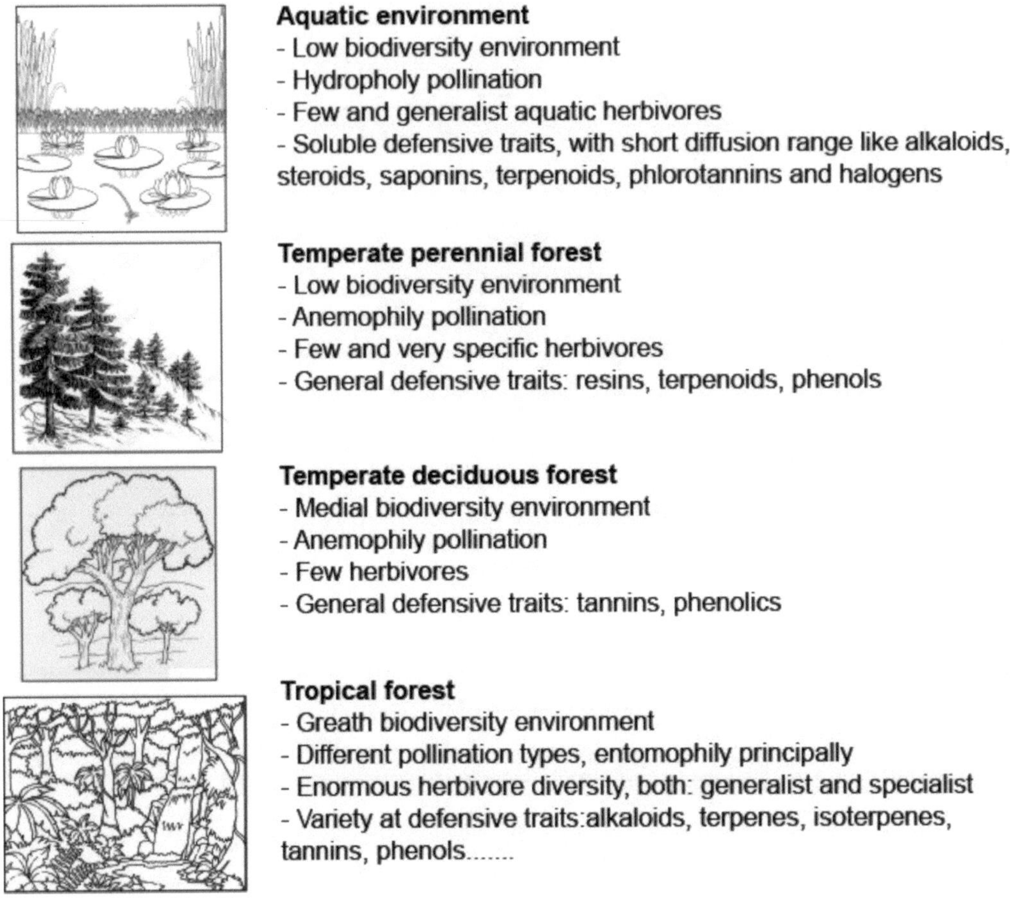

Figure 2. Characteristics related to defense mechanisms in plants of different environments.

In a temperate deciduous forest at Powdermill Nature Reserve, the leaf damage caused by herbivores and in majority of individual had a low rate than 2% that can be due to low herbivore densities and poor degrees of specialization thereof [64].

Also, in stressful environment like high-elevation alpine plant communities with low temperatures, plant species have asexual reproduction by rhizomes resulting in clones. Clonal species have developed a tolerance strategy against herbivores and reduced investment in chemical defense [65].

3.2. Tropical forest

The most recent plant community is the tropical forest (see distribution in **Figure 1**), which originates toward of the end of Cretaceous period when angiosperms take over the plant diversity [66]. In general, plants at tropical forest have a great variety of chemical compounds for defense, principally blends (**Figure 2**, **Table 1**).

In tropics, there is a high herbivory variation explained by multiple syndromes in plant defense strategies, driven by leaf nutritional quality, in relation to nutrition defense [67], where most extreme plants can combine high chemical defense, low nutritional quality and asynchronous leaf expansion, reduces to minimal its vulnerability [68].

At tropical forest, chemical plant defenses have diverged recently and increased their diversity, because there is a high herbivore pressure due to high insect diversity [69]. Tropical forest may hold more than 650 tree species per hectare, in that species interact each, and pests may promote plant diversity including that in leaves of a unique tropical tree, there are hundreds of different chemical defensive compounds. In addition, herbivore diversity and abundance, rates of herbivory, and host specificity are higher in the tropical than temperate plants (see **Table 1** for comparison) [70]. For example, in Amazonian forest canopy, there are concentrations of one to two orders of magnitude in value of foliar phenols, lignin, and cellulose [71].

Tropical forest has been considering an unproductive habitat where plants need investment in defensive traits because they cannot utilize molecules and energy simultaneously to defense, growth, and replacing loss tissues [6, 45, 72]. Then, synergistic interactions among various defensive traits offer an effective resistance, which is reflected with an increase in the simultaneous expression for direct and indirect defenses [73]. When plants exceed the capacity to store, constitutive secondary metabolites could avoid autotoxicity [74].

Mixtures of defensive compounds allow plant increase resistance, including attack from new herbivore-related congeners, considering that species interactions are stronger in tropics [75].

In general, plant species in tropical forest have a high defensive diversity, in which plant species are chemically unique in their communities [76]. It should be noted that chemical compounds implicate in defensive traits, and different interactions between molecules to perform defenses are equally distributed at family, genus, and species level [71].

For example, in tree genus *Inga*, there are a great variety of defensive traits, like phenolics, which includes polygalloylated compounds, polymers of flavan-3-ols with different substitutions, triterpene saponins, and the amino acid tyrosine. Moreover, plant can identify the agent, amount, and timing of damage and produce a particular induced response, and its response differs in low- and high-risk environments [77].

Amino acid tyrosine can be redirected into other primary and secondary metabolites, and its accumulation in excess in young leaves may not be adaptive as they would persist once the leaf was full size and protected by toughness [78].

Another genus well characterized about its defensive chemical compounds is *Piper*, which is broadly represented at tropical forest in the world [79]. The most bioactive compounds reported by *Piper* are amides, a group nitrogen-based compounds stored at leaves and fruits to defend that genera against herbivores [80]. In *Piper*, prenylated benzonic acid, chromene, and dimeric chromane at concentrations higher than 10% of dry weight of leaf material that compounds have synergistic or additive effect against herbivore attack also have been reported. In addition, concentration of these metabolites is correlated with increasing elevation in relation with UV exposure and photoactive properties, and more toxic plants support a lower diversity of specialist herbivores [81].

Sometimes, plant species can response locally to different herbivores, as *Datura stramonium*: the plant can be eaten by generalist and specialist herbivores at great geographic range and produces the alkaloid atropine and its derivate, less toxic, scopolamine. The secondary plant compound is more effective against herbivores specialist, but the precursor is still effective against generalist. Then, when there is a community of generalists, *D. stramonium* produces atropine [82].

Another example is *Zamia stevensonii*, which produces azoxy glycosides (AZGs), highly toxics with mutagenic and carcinogenic properties. AZGs are an excellent defense against generalist herbivores, but are not sufficient to specialist [83].

Another important group of chemical compounds, relevant in defensive plant traits, is phenols, including tannins, which at media concentration reduces herbivory, through reduction of digestibility of plant tissues, and increases immune responses [84].

Terpenes also are present at plants in tropical forest; in that, these compounds protect against abiotic factors such as light, heat, and drought and against herbivores. In Borneo rain forest, foliar terpene presence in 73 of 75 plant species has been analyzed (97%), 15 monoterpenes and 65 sesquiterpenes. This suggests that terpenes can be a favorable selective trait in rainforest [46].

Currently 25,000 structures of terpenes approximately have been reported; some of them are volatile and can be synthesized de novo or are stored in leaves, stems, and trunks and are released in response to attack [85]. One plant can release a highly complex blend, which can include up to 200 volatile terpenes, and its effect is due to direct toxicity, repulsion to herbivores, or attraction of herbivore enemies [85, 86].

Among defensive traits in tropical forest, some are strongly correlated with herbivore damage: leaf size, shearing resistance, cellulose, and ash content. Then, large leaves are more susceptible to herbivory. Other three factors—shear toughness, cellulose content, and ash, which is a mixture of calcium oxalates and phytoliths—reduce herbivore damage acting as structural defenses. These strategies are very efficient and have a relatively low energetic cost [45].

Interestingly, lianas have increased cover and abundance. That plant forms are genetically predisposed to reduce structure and defense traits for investment more in chemical implicates in growth and light capture, wherewith lianas response to stress conditions, like warmer and drier conditions [87].

At tropical forest are common indirect defenses to reduce herbivore attack. In that way, plants provide house, nourish or attract organisms like ants or parasitoids [88], by production of refuges or nesting sites, extrafloral nectar, food bodies or/ and volatile compounds (VOCs) [89].

For example, extrafloral nectar production increases in herbivory and diminishes in the herbivore absence, because that is the secret to attract predators like ants, who defend their food sources and parasitoids. Extrafloral nectar consists in sugars, proteins, lipids, mineral nutriments, and antioxidants and can attract organisms like mites, ladybird beetles, wasp, lacewing larvae, and spiders [90, 91].

VOCs also attract other organisms to improve defense, can attract pollinators, repel herbivores, and are used by plants for communication among them [92, 93] to alert of a possible future attack [94].

4. Conclusions

Every ambient has their own biotic and abiotic selective pressures, and plants are able to respond differentially. In general, there are a great variety of defensive traits in plants, and they are different at distinct environments. In aquatic environment, the principal defensive traits must be water soluble and diffusible, but their action time and range are short, because aquatic plants need to produce this constantly. For aquatic plants, biomass production is more relevant than defensive traits outlay, then they prefer investment in growth and photosynthesis.

In terrestrial environments, plants need to accumulate defensive compounds for an eventual attack, but the quantity depends on biomass replacement rate. If ambient conditions allow rapid biomass formation, plant accumulates less defensive compounds; in contrast, slow accumulation in biomass induces the great defensive compound accumulation. Because at temperate forest, plants accumulate greater quantity of defensive compounds, particularly resins and phenolics.

Herbivore diversity and pathogens also contribute in plant chemical production. In places, with high herbivore diversity, plants produce a plethora of compounds for defense so much for generalist such as specialist; then at tropical forest, there are a greater variety of chemical defensive compounds, especially complex chemical mixtures.

Author details

Alina Morquecho-Contreras[1]*, Carmen Zepeda-Gómez[2] and Hermilo Sánchez-Sánchez[2]

*Address all correspondence to: alimorcon@gmail.com

1 Agricultural Faculty, Autonomous University of Mexico State, Toluca, Mexico

2 Science Faculty, Autonomous University of Mexico State, Toluca, Mexico

References

[1] Cimino G, Ghiselin MT. Marine natural products chemistry as evolutionary narrative, In: M.T., McClintock JB, Baker BJ. Editors. Marine Chemical Ecology. Boca Raton, FL, USA: CRC Press; 2001. pp. 115-154

[2] Jackrel SL, Wootton T. Cascading effects of induced terrestrial plant defenses on aquatic and terrestrial ecosystem function. Proceedings of Royal Society B. 2015;**282**:20142522. doi: 10.1098/rspb.2014.2522

[3] Andereg WRL, Jeffrey AH, Fisher RA, et al. Tree mortality from drought, insects, and their interactions in a changing climate. New Phytologist. 2015;**208**:674-683. DOI: 10.1111/nph.13477

[4] Verhage A, SCM v W, CMJ P. Plant immunity: It's the hormones talking, but what do they say? Plant Physiology. 2010;**154**:536-540. DOI: 10.1104/pp.110.161570

[5] War AR, Paulraj MG, Ahmad T, Buhroo AA, Hussain B, Ignacimuthu S, Sharma HC. Mechanisms of plant defense against insect herbivores. Plant Signaling and Behaviour. 2012;**7**:1306-1320. DOI: 10.4161/psb.21663

[6] Coley PD. Interspecific variation in plant anti-herbivore properties: The role of habitat quality and rate of disturbance. New Phytologist. 1987;**106**:251-256

[7] Hay ME. Marine terrestrial contrasts in the ecology of plant chemical defenses against herbivores. Tree. 1991;**6**:17

[8] Johnson MTJ. Evolutionary ecology of plant defences against herbivores. Functional Ecology. 2011;**25**:305-311

[9] Gong B, Zhang G. Interactions between plants and herbivores: A review of plant defense. Acta Ecologica Sinica. 2014;**34**:325-336

[10] Whittaker RH, Feeny PP. Allelochemics: Chemical interactions between species. Science. 1971;**171**:757-770

[11] Wood KA, O'Hare MT, McDonald C, Shearle KR, Daunt F, Stillman RA. Herbivores regulation of plant abundance in aquatic ecosystems. Biologica Reviews. 2016;**92**:1128-1141. DOI: 10.1111/brv.12272

[12] Maschek JA, Baker BJ. The chemistry of algal secondary metabolism. In: Amsler CH. Editor. Algal Chemical Ecology. Berlin: Springer; 2008. pp. 1-20

[13] Hay ME, Steinberg PD. The chemical ecology of plant-herbivore interactions in marine versus terrestrial communities. In: Rosenthal J, Berenbaum M. editors. Herbivores: Their Interaction with Secondary Plant Metabolites. Evolutionary and Ecological Processes. vol. II. New York: Academic Press; 1999. pp. 371-413

[14] Pereira RC, Yoneshigue-Valentin Y. The role of polyphenols from the tropical brown alga *Sargassum furcatum* on the feeding by amphipod herbivores. Botanica Marina. 1999;**42**:441-448

[15] Li YX, Wijesekara I, Li Y, Kim SK. Phlorotannins as bioactive agents from brown algae. Process Biochemistry. 2011;**46**:2219-2224

[16] Stiger-Pouvreau V, Jégou C, Cérantola S, Guérard F, Lann KL. Phlorotannins in Sargassaceae Species from Brittany (France): Interesting molecules for ecophysiological and valorisation purposes. In: Jacquot JP, Gadal P, serial editors. and Bourgougnon N, serial volume editor. Advances in Botanical Research. Vol. 71. Sea plants. New York: Academic Press, Elsevier Ltd. 2014. pp 379-412

[17] Blunt JW, Copp BR, Hu WP, Munro MHG, Northcote PT, Prinsep MR. Marine natural products. Natural Products Reports. 2009;**26**:170-244

[18] Hay ME, Fenical W. Marine plant herbivore interactions, the ecology of chemical defense. Annual Review of Ecology and Systematics. 1988;**19**:111-145

[19] Wolfe GV. The chemical defense ecology of marine unicellular plankton: Constrains, mechanisms, and impacts. Biological Bulletin. 2000;**918**:225-244

[20] Connan S, Goulard F, Stiger V, Deslandes E, Ar GE. Interspecific and temporal variation in phlorotannin levels in an assemblage of brown algae. Botanica Marina. 2004;**47**:410-416

[21] Subhashini P, Dilipan E, Thangaradjou T, Papenbrock J. Bioactive natural products from marine angiosperms: Abundance and functions. Natural Products Bioprospectives. 2013;**3**:129-136

[22] Lodge DM. Herbivory on freshwater macrophytes. Aquatic Botany. 1991;**41**:195-224

[23] Newman RM. Herbivory and detritivory on fresh-water macrophytes by invertebrates: A review. Journal of the North American Benthological Society. 1991;**10**:89-114

[24] Hutchinson GE. A Treatise on Limnology Volume III. Limnological Botany. New York: John Wiley and Sons; 1975

[25] Wetzel RG. Limnology. Fort Worth, TX: Saunders College Publishing; 1983

[26] Lodge DM, Cronin G, Van Donk E, Froelich AJ. Impact of herbivory on plant standing crop: Comparisons among biomes, between vascular and nonvascular plants, and among freshwater herbivore taxa. In: Jeppesen E, Søndergaard M, Søndergaard M, Christoffersen K, editors. In the Structuring Role of Submerged Macrophytes in Lakes. New York: Springer-Verlag; 1998. pp. 149-174

[27] Bakker ES, Pages JF, Arthur R, Alcoverro T. Assessing the role of large herbivores in the structuring and functioning of freshwater and marine angiosperm ecosystems. Ecography. 2016;**39**:162-179

[28] Sondergaard M, Bruun L, Lauridsen T, Jeppensen E, Madsen TV. The impact of grazing waterfowl on submerged macrophytes: In situ experiments in a shallow eutrophic lake. Aquatic Botany. 1996;**53**:73-84

[29] Newman RM, Kerfoot WC, Hanscom Z. Watercress allelochemical defends high-nitrogen foliage against consumption: Effects on freshwater invertebrate herbivores. Ecology. 1996;**77**:2312-2323

[30] Cronin G, Lodge DM, Hay ME, Miller M, Hill AM, Horvath T, Bolser RC, Lindquist N, Wahl M. Crayfish feeding preferences for fresh water macrophytes: The influence of plant structure and chemistry. Journal of Crustacean Biology. 2002;**22**:708-718

[31] Prusak AC, O'Neal J, Kubanek J. Prevalence of chemical defenses among freshwater macrophytes. Journal of Chemical Ecology. 2005;**31**:1145-1160

[32] Cronk JK, Fennessy MS. Wetland Plants: Biology and Ecology. Boca Raton, Florida: Lewis Publishers; 2001

[33] Su KL, Staba EJ, Hajj YA. Preliminary chemical studies of aquatic plants from Minnesota. Lloydia. 1973;**36**:72-79

[34] Ostrofsky ML, Zettler ER. Chemical defenses in aquatic plants. Journal of Ecology. 1986;**74**:279-287

[35] Erhard D, Pohnert G, Gross EM. Chemical defense in *Elodea nuttallii* reduces feeding and growth of aquatic herbivorous Lepidoptera. Journal of Chemical Ecology. 2007;**33**:1646-1661

[36] Gross EM, Bakker ES. The role of plant secondary metabolites in freshwater macrophyte herbivore interactions: Limited or unexplored chemical defences?. In: Iason GR, Dicke M, Hartley SE. Editors. The Ecology of Plant Secondary Metabolites: From Genes to Global Processes. New York: Cambridge University Press. 2012

[37] Smolders AJP, Vergeer LHT, Van der Velde G, Roelofs JGM. Phenolic contents of submerged, emergent and floating leaves of aquatic and semi-aquatic macrophyte species: Why do they differ? Oikos. 2000;**91**:307-310

[38] Cronin G, Wissing KD, Lodge DM. Comparative feeding selectivity of herbivorous insects on water lilies: Aquatic vs. semiterrestrial insects and submersed vs. floating leaves. Freshwater Biology. 1998;**39**:243-257

[39] Bolser RC, Hay ME. A field test of inducible resistance to specialist and generalist herbivores using the water lily *Nuphar luteum*. Oecologia. 1998;**116**:143-153

[40] Fornoff F, Gross EM. Induced defense mechanisms in an aquatic angiosperm to insect herbivory. Oecologia. 2014;**175**:173-185. DOI: 10.1007/s00442-013-2880-8

[41] Hori K, Maruyama F, Otha H. *Klebsormidium flaccidum* genome reveals primary factors for plant terrestrial adaptation. Nature Communication. 2014;**5**:3978. DOI: 10.1038/ncomms49

[42] Shurin JB, Gruner DS, Hillebrand H. All wet or dried up? Real differences between aquatic and terrestrial food weebs. Proceedings of Royal Society B. 2006;**273**:1-9. DOI: 10.1098/rspb.2005.3377

[43] Abdala-Roberts L, Rasmann S, Benny-Mier y Terán JC, Covelo F, Glauser G, Moreira X. Biotic and abiotic factors associated with altitudinal variation in plant traits and herbivory in a dominant oak species. American Journal of Botany. 2016;**103**:1-9. DOI: 10.3732/ajb.1600310

[44] Herms DA, Mattson WJ. The dilemma of plants: To grow or defend. Quarterly Review of Biology. 1992;**67**:283-335

[45] Cardenas RE, Valencia R, Kraft NJB, Argoti A, Dangles O. Plant traits predict inter- and intraspecific variation in susceptibility to herbivory in a hyperdiverse Neotropical rain forest tree community. Journal of Ecology. 2014;**102**:939-952. DOI: 10.1111/1365-2745.12255

[46] Sardans J, Llusia J, Owen SM, Ninnemets Ü, Peñuelas J. Screening study of leaf terpene concentration of 75 borneo rainforest plant species: Relationships with leaf elemental concentrations and morphology. Records of Natural Products. 2015;**9**:19-40

[47] Nystedt B, Street NR, Wetterbom A, Zuccolo A, Yao-Cheng, Douglas L, Scofield G, et al. The Norway spruce genome sequence and conifer genome evolution. Nature. 2013;**497**:579-584. DOI: 10.1038/nature1221

[48] Turcotte MM, Davies TJ, Thomsen CJM, Johnson MTJ. Macroecological and macroevolutionary patterns of leaf herbivory across vascular plants. Proceedings of Royal Society B. 2014;**281**:20140555. DOI: 10.1098/rspb.2014.0555

[49] Stiegel S, Entling MH, Mantilla-Contreras J. Reading the leaves' palm: Leaf traits and herbivory along the microclimatic gradient of forest layers. PLoS ONE. 2017;**12**:e0169741. DOI: 10.1371/journal.pone.0169741

[50] Franceschi VR, Krokene P, Christiansen E, Krekling T. Anatomical and chemical defenses of conifer bark against bark beetles and other pests. New Phytologist. 2005;**167**: 353-376

[51] Ferrenberg S, Kane JM, Mitton B. Resin duct characteristics associated with tree resistance to bark beetles across lodgepole and limber pines. Oecologia. 2013;**174**:1283-1292. DOI: 10.1007/s00442-013-2841-2

[52] Eyles A, Bonello P, Ganley R, Mohammed C. Induced resistance to pest and pathogens in trees. New Phytologist. 2010;**185**:893-908

[53] Franceschi VR, Nakata PA. Calcium oxalate in plants: Formation and function. Annual Review of Plant Biology. 2005;**56**:41-71

[54] Hudgins JW, Christiansen E, Franceschi VR. Methyl jasmonate induces changes mimicking anatomical and chemical defenses in diverse members of the Pinaceae. Tree Physiology. 2003;**23**:361-371

[55] Hudgins JW, Christiansen E, Franceschi V. Induction of anatomically based defense responses in stems of diverse conifers by methyl jasmonate: A phylogenetic perspective. Tree Physiology. 2004;**24**:251-264

[56] Ruel JJ, Ayres MP, Lorio PL Jr. Loblolly pine responds to mechanical wounding with increased resin flow. Canadian Journal of Forest Research. 1998;**28**:596-602

[57] Pan HF, Lundgren LN. Phenolic extractives from root bark of Picea abies. Phytochemistry. 1995;**39**:1423-1428

[58] Pan HF, Lundgren LN. Phenolics from inner bark of *Pinus sylvestris*. Phytochemistry. 1996;**42**:1185-1189

[59] Viiri H, Annila E, Kitunen V, Niemela P. Induced responses in stilbenes and terpenes in fertilized Norway spruce after inoculation with blue-stain fungus, *Ceratocystis polonica*. Trees – Structure and Function. 2001;**15**:112-122

[60] Klepzig KD, Kruger EL, Smalley EB, Raffa KF. Effects of biotic and abiotic stress on induced accumulation of terpenes and phenolics in red pines inoculated with bark beetle-vectored fungus. Journal of Chemical Ecology. 1995;**21**:601-626

[61] Franceschi VR, Krekling T, Berryman AA, Christiansen E. Specialized phloem parenchyma cells in Norway spruce (Pinaceae) bark are an important site of defense reactions. American Journal of Botany. 1998;**85**:601-615

[62] Hietala AM, Kvaalen H, Schmidt A, Jøhnk N, Solheim H, Fossdal CG. Temporal and spatial profiles of chitinase expression by Norway spruce in response to bark colonization by *Heterobasidion annosum*. Appl Environ Microbiol. 2004;**70**:3948-53. DOI: 10.1128/AEM.70.7.3948-3953.2004

[63] Xiao-Fei W, Jian-Feng L, Wen-Qiang G, et al. Defense pattern of Chinese cork oak across latitudinal gradients: Influences of ontogeny, herbivory, climate and soil nutrients. Scientific Reports. 2016;**6**:27269. DOI: 10.1038/srep27269

[64] Murphy SJ, Xu K, Comita LS. Tree seedling richness, but not neighborhood composition, influences insect herbivory in a temperate deciduous forest community. Ecology and Evolution. 2016;**6**:6310-6319. DOI: 10.1002/ece3.2336

[65] Pelliseier L, Litsios G, Fishbein M, Salamin N, Agrawal AA, Rasmann S. Different rates of defense evolution and niche preferences inclonal and nonclonal milkweeds (*Asclepias* spp.). New Phytologist. 2016;**209**:1230-1239. DOI: 10.1111/nph.13649

[66] Maley J. The African rain forest—main characteristics of changes in vegetation and climate from the upper cretaceous to the quaternary. Proceedings of the Royal Society of Edinburgh, Section B: Biological Sciences. 1996;**104**:31-73. DOI: 10.1017/S0269727000006114

[67] Agrawal AA, Fishbein M. Plant defense syndromes. Ecology. 2006;**87**:S132-S149

[68] Kursar TA, Coley PD. Convergence in defense syndromes of young leaves in tropical rainforests. Biochemical Systematics and Ecology. 2003;**31**:929-949

[69] Salazar D, Marquis RJ. Herbivore pressure increases toward the equator. Proceedings of National Academy of Sciences. 2012;**109**:12616-12620

[70] Coley PD, Kusar TA. On tropical forest and their pests. Science. 2014;**343**:35-36. DOI: 10.1126/science.1248110

[71] Anser GP, Martin RE, Tupayachi R, Anderson CB, Sinca F, Carranza-Jiménez L, Martinez P. Amazonian functional diversity from forest canopy chemical assembly. PNAS. 2014;**111**:5604-5609. DOI: 10.1073/pnas.1401181111

[72] Janzen DH. Tropical blackwater rivers, animals, and mast fruiting by the Dipterocarpaceae. Biotropica. 1974;**6**:69-103

[73] Pellissier L, Moreira X, Danner H, Serrano M, Salamin N, van Dam NM, Rasmann S. The simultaneous inducibility of phytochemicals related to plant direct and indirect defences against herbivores is stronger at low elevation. Journal of Ecology. 2016;**104**:1116-1125. DOI: 10.1111/1365-2745.12580

[74] Agrawal A, Karban R. Why induced defenses may be favored over constitutive. Pp 45-65 *in* Tollrian R, Harvell CD. The Ecology and Evolution of Inducible Defenses. 1999. Princeton: Princeton University Press. DOI: 10.1234/12345678

[75] Becerra JX. On the factors that promote the diversity of herbivorous insects and plants in tropical forests. PNAS. 2015;**112**:6098-6103. DOI: 10.1073/pnas.1418643112

[76] Becerra JX. Macroevolutionary and geographical intensification of chemical defense in plants driven by insect herbivore selection pressure. Current Opinion in Insect Science. 2015;**8**:1-7. DOI: 10.1016/j.cois.2015.01.012

[77] Bixenmann RJ, Coley PD, Weinhold A, Kursar TA. High herbivore pressure favor constitutive over induced defense. Ecology and Evolution. 2016;**6**:6037-49. DOI: 10.1002/ece 3.2208

[78] Lucas PW, Turner IM, Dominy NJ, Yamashita N. Mechanical defences to herbivory. Annals of Botany 2000;**86**:913-920. DOI: 10.1006/anbo.2000.1261

[79] Salazar D, Jaramillo AM, Marquis RJ. Chemical similarity and local community assembly in the species rich tropical genus Piper. Ecology. 2016;**97**:3176-3183. DOI: 10.1002/ecy.1536

[80] Whitehead SR, Obando Quesada MF, Bowers MD. Chemical tradeoffs in seed dispersal: Defensive metabolites in fruits deter consumption by mutualist bats. Oikos. 2015;**125**: 927-937. DOI: 10.5061/dryad.br022

[81] Glassmire AE, Jeffrey Cs, Forister ML, et al. Intraspecific phytochemical variation shapes community and population structure for specialist caterpillars. New Phytologist. 2016;**212**:208-219. DOI: 10.1111/nph.14038

[82] Castillo G, Cruz LL, Tapia-López R, Olmedo-Vicente E, Carmona D, Anaya-Lang AL, Fornoni J, Andraca-Gómez G, Valverde PL, Nuñez-Farfán J. Selection mosaic exerted by specialist and generalist herbivores on chemical and physical defense of *Datura stramonium*. PLoS ONE. 2014;**9**:e102478. DOI: 10.1371/journal.pone.0102478

[83] Prado A, Sierra A, Windsor D, Bede JC. Leaf traits and herbivory level in a tropical gymnosperm, *Zamia stevensonii* (Zamiaceae). American Journal of Botany. 2014;**101**:437-447. DOI: 10.3732/ajb.1300337

[84] López-Carretero A, Boege K, Díaz-Castelazo C, Domínguez Z, Rico-Gray V. Influence of plant resistance traits in selectiveness and species strength in a tropical plant-herbivore network. American Journal of Botany. 2016;**103**:1-13. DOI: 10.3732/ajb.1600045

[85] Courtois EA, Dexter KG, Paine CET, Stien D, Engel J, Baraloto C, Chave J. Evolutionary patterns of volatile terpene emissions across 202 tropical tree species. Ecology and Evolution. 2016;**6**:2854-2864. DOI: 10.1002/ece3.1810

[86] Dicke M, JJA v L. Multitrophic effects ofherbivore-induced plant volatiles in an evolutionary context. Entomologia Experimentalis et Applicata. 2000;**97**:237-249

[87] Asner GP, Martin RE. Contrasting leaf chemical traits in tropical lianas and trees: Implications for future forest composition. Ecological Letter. 2012;**15**:1001-1007

[88] Fürstenberg-Hägg J, Zagrobelny M, Bak S. Plant defense against insect herbivores. International Journal of Molecular Science. 2013;**14**:10242-10297

[89] Dicke M, Sabelis MW. How plants obtain predatory mites as bodyguards. Netherlands Journal of Zoology. 1988;**38**:149-165

[90] Wäckers FL, Zuber D, Wunderlin R, Keller F. The effect of herbivory on temporal and spatial dynamics of foliar nectar production in cotton and castor. Annals of Botany. 2001;**87**:365-370

[91] Heil M. Indirect defence via tritrophic interactions. New Phytologist Journal. 2007; **178**:41-61

[92] Kessler A, Baldwin IT. Defensive function of herbivore-induced plant volatile emissions in nature. Science. 2001;**291**:2141-2144

[93] Dudareva N, Negre F, Nagegowda DA, Orlova I. Plant volatiles: Recent advances and future perspectives. Critical Reviews in Plant Science. 2006;**25**:417-440

[94] Engelberth J, Alborn HT, Schmelz EA, Tumlinson JH. Airborne signals prime plants against insect herbivore attack. Proceedings of National Academy of Sciences of the United States of America. 2004;**101**:1781-1785

Chapter 6

Vachellia (*Acacia*) *karroo* Communities in South Africa: An Overview

Mamokete Dingaan and Pieter J. du Preez

Additional information is available at the end of the chapter

http://dx.doi.org/10.5772/intechopen.70456

Abstract

Vachellia karroo is a useful and widespread tree in Africa. It belongs to the family Fabaceae, which is the third largest woody plant family in southern Africa. This is an ecologically and economically important species as almost all of its parts, including bark, pods, seeds, leaves and thorns, are extremely useful to both humans and animals. Various commercial products are also obtained from the tree, and gum is one of the most important products. *V. karroo* in South Africa has an extensive distribution range that includes several biomes. It is very adaptable and has wide habitat tolerance, growing under many differing conditions of soil, climate, and altitude. Although it is often associated with heavy, clayey soils on the banks of rivers and streams, it also grows in bushveld, dry thornveld, grassland and woodland. *V. karroo* is easy to grow and as a result can become an aggressive invader of valuable farming land and grazing areas, a phenomenon usually referred to as bush encroachment. An analysis of historic data comprising 1553 relevés and 2006 species, compiled from all areas of South Africa where *V. karroo* is known to occur was conducted, and TWINSPAN classification produced five main vegetation types.

Keywords: *Acacia*, biome, bush encroachment, ecological significance, economic value, geographical range, soil enrichment, sweet thorn, *Vachellia*

1. Introduction

Vachellia karroo is a highly useful tree that is widespread throughout Africa [1], and it is the most widely distributed tree in South Africa [2]. It belongs to the family Fabaceae (Legume family), which is one of the largest woody plant families in southern Africa. Species of the *Vachellia* genus vary in their distribution range; there are species that are very widely distributed and occupy a diverse range of habitats, while others have a very restricted distribution [3]. The species are a prominent feature in the Savanna biome (bushveld) in South Africa but can also form local dominant stands in other biomes such as the Grassland and Nama-Karoo biomes. Those with a broad distribution range, like *V. karroo,* occur in several biomes [1].

The *Vachellia* species are pod-bearing woody plants that range from shrubs to large trees. They can be sprawling or climbing, and this character differs with habitat [4]. This genus in Africa is readily recognised by its thorns, which are typically paired and straight. These thorns are modified stipules, which become hard and spiny [1, 5] and are important for identification of the trees [6]. *Vachellia* trees can further be distinguished by their characteristic growth form, by bark, and also by pods. This is however a taxonomically difficult genus containing a number of closely related species whose recognition and identification are not always simple [3, 5].

1.1. *Vachellia* split from the *Acacia* genus

Until 2005, *V. karroo* was known as *Acacia karroo,* but according to recent taxonomic research and molecular evidence, the *Acacia* genus was shown to be polyphyletic [7]. It could not be maintained as a single entity, and a proposal was put forward for it to be divided into five genera [8–10]. According to the new proposed classification, ratified at the International Botanical Congress in Vienna in July 2005, *Acacia* genus was split into five monophyletic genera, with all the African *Acacia* now falling under *Vachellia* and *Senegalia* as follows:

(i) *Acacia,* preserved for more than 960 largely Australian species, which all belonged to the former sub-genus *Phyllodineae.*

(ii) *Vachellia,* former sub-genus *Acacia,* approximately 161 pantropical species (Africa, Asia and Latin America).

(iii) *Senegalia,* former sub-genus *Aculeiferum,* with 203 pantropical species (Africa, Asia and Latin America).

(iv) *Acaciella,* former sub-genus *Aculeiferum* section *Filicinae,* contains 15 species from the Americas.

(v) A yet unnamed genus with 13 species from the Americas.

There were objections toward preserving the name *Acacia* for the Australian and other related species [11], but the decision taken in Vienna in 2005 was finalised at the next International Botanical Congress held in Melbourne in 2011. Before the split, there was a total of 40 *Acacia* species, subspecies and varieties represented in South Africa [1]. The split has now resulted in 23 species, subspecies and varieties of *Vachellia* and 17 of *Senegalia* (**Table 1**). The key diagnostic character distinguishing *Vachellia* from *Senegalia* is the presence of stipular spines in *Vachellia,* while *Senegalia* may have prickles but always lack stipular spines [12].

Vachellia		***Senegalia***	
Old name	**New name**	**Old name**	**New name**
Acacia borleae	*Vachellia borleae*	*Acacia ataxacantha*	*Senegalia ataxacantha*
Acacia davyi	*Vachellia davyi*	*Acacia brevispica* subsp. *dregeana*	*Senegalia brevispica* subsp. *dregeana*
Acacia erioloba	*Vachellia erioloba*	*Acacia burkei*	*Senegalia burkei*
Acacia exuvialis	*Vachellia exuvialis*	*Acacia caffra*	*Senegalia caffra*
Acacia gerrardii var. *gerrardii*	*Vachellia gerrardii* var. *gerrardii*	*Acacia erubescens*	*Senegalia erubescens*
Acacia grandicornuta	*Vachellia grandicornuta*	*Acacia fleckii* = *A. cinerea*	*Senegalia cinerea*
Acacia haematoxylon	*Vachellia haematoxylon*	*Acacia galpinii*	*Senegalia galpinii*
Acacia hebeclada subsp. *hebeclada*	*Vachellia hebeclada* subsp. *hebeclada*	*Acacia goetzei* subsp. *microphylla*	*Senegalia goetzei* subsp. *microphylla*
Acacia karroo	*Vachellia karroo*	*Acacia hereroensis*	*Senegalia hereroensis*
Acacia luederitzii var. *luederitzii*	*Vachellia luederitzii* var. *luederitzii*	*Acacia kraussiana*	*Senegalia kraussiana*
Acacia luederitzii var. *retinens*	*Vachellia luederitzii* var. *retinens*	*Acacia mellifera* subsp. *detinens*	*Senegalia mellifera* subsp. *detinens*
Acacia nebrownii	*Vachellia nebrownii*	*Acacia nigrescens*	*Senegalia nigrescens*
Acacia nilotica subsp. *kraussiana*	*Vachellia nilotica* subsp. *kraussiana*	*Acacia polyacantha* subsp. *campylacantha*	*Senegalia polyacantha* subsp. *campylacantha*
Acacia permixta	*Vachellia permixta*	*Acacia schweinfurthii* var. *schweinfurthii*	*Senegalia schweinfurthii* var. *schweinfurthii*
Acacia rehmanniana	*Vachellia rehmanniana*	*Acacia senegal* var. *leiorhachis*	*Senegalia senegal* var. *leiorhachis*
Acacia robusta subsp. *clavigera*	*Vachellia robusta* subsp. *clavigera*	*Acacia senegal var. rostrata*	*Senegalia senegal* var. *rostrata*
Acacia robusta subsp. *robusta*	*Vachellia robusta* subsp. *robusta*	*Acacia welwitschii* subsp. *delagoensis*	*Senegalia welwitschii* subsp. *delagoensis*
Acacia sieberiana var. woodii	*Vachellia sieberiana* var. *woodii*		
Acacia stuhlmannii	*Vachellia stuhlmannii*		
Acacia swazica	*Vachellia swazica*		
Acacia tenuispina	*Vachellia tenuispina*		
Acacia tortilis subsp. *heteracantha*	*Vachellia tortilis* subsp. *heteracantha*		
Acacia xanthophloea	*Vachellia xanthophloea*		

Table 1. South African *Vachellias (Acacias)* and their new name combinations [1, 12, 13].

1.2. Morphological variation of *V. karroo*

The species displays considerable variation in its appearance, size and other characters [3, 14]. This variation in *V. karroo* is seemingly regional with plants from different geographical areas appearing distinctly different with regard to one or more features [1, 13, 15]. The "typical" form of *V. karroo* grows in the Karoo, Free State, KwaZulu-Natal, and some northern parts of the country [1]. It is a small to medium-sized tree commonly growing to 5–12 m in height but may become a very large tree of up to 22 m on river banks or in other favourable conditions [1, 2]. The tree is usually single-stemmed though sometimes multi-stemmed, branching high above the ground to give a rounded crown (**Figure 1**).

The typical *V. karroo* has a rough, longitudinally fissured bark which is dark on the trunk (**Figure 2A**) and main branches but rusty red in younger branches (**Figure 2B**). The foliage is generally dense and comprises dark green compound leaves (**Figure 2C**). Inflorescences are balls of small sweetly scented yellow flowers (**Figure 2D**), while the pods are flat, mostly sickle shaped with minor constrictions between seeds and dehiscent (**Figure 2E**). The thorns are long, paired, straight, and shining white (**Figure 2F**) and indicate an adaptation of *V. karroo* to its environment because of their protective function [2]. They are larger and abundant on the lower branches that are within reach of animals (and also on young trees) (**Figure 3**) but fewer on the higher parts of larger (and old) trees [2, 4].

Figure 1. A *Vachellia karroo* tree near Bloemfontein, South Africa (photo: M. Dingaan).

Figure 2. *Vachellia karroo* trunk (A), branches (B), leaves (C), flowers (D), pods (E) and thorns (F) (photos: M. Dingaan).

Due to the extreme variation in *V. karroo* form, many of the variations have been described as different species in the past, resulting in numerous synonyms. The differences in form have thus been considered by some botanists to be distinct enough to warrant division of the species into sub-species or at least varieties or even to again regard some forms as different species altogether. Ross [13] concluded that it would be preferable to regard *V. karroo*

Figure 3. A young *Vachellia karroo* tree, splendidly armoured with long white thorns (photo: M. Dingaan).

Morphological variation [1, 13]	**Current taxonomic status [5, 16]**
1. White-barked trees or shrubs with short spines, found in Eastern Cape, KwaZulu-Natal, Mpumalanga and neighbouring countries (Swaziland, Zimbabwe and Mozambique)	*V. natalitia*
2. Small slender shrubs found near the Kei River mouth, Eastern Cape	*V. dyeri*
3. Fire resistant shrubs in the Nongoma District, KwaZulu-Natal	Formerly *A. inconflagrabilis*, still a synonym
4. Slender sparsely branched trees in the Hluhluwe and Umfolozi Game Reserves, KwaZulu-Natal	*V. theronii* (or *V. montana*)
5. Large trees with greyish-white bark along the Tugela River mouth (KwaZulu-Natal), and northwards into Mozambique	*V. kosiensis*
6. Sparse indumentum on young shoots, leaves peduncles and pods on the Highveld from Pretoria eastwards (for example Sekhukhuneland, Limpopo)	*V. robbertsei*
7. Small shrubby form on the Springbok flats north of Pretoria	Still *V. karroo*, closely resembles *V. tenuispina*

Table 2. The *Vachellia karroo* complex.

as a variable polymorphic species rather than to divide the species into a number of infra-specific taxa. Regardless, the *V. karroo* complex has recently been split, with some authors recognising the following as distinct species (**Table 2**): *V. natalitia, V. dyeri, V. kosiensis* [5, 16] and *V. theronii* (previously published incorrectly as *V. montana,* i.e., an invalid name) [17]. Coates Palgrave [5] further recognises *V. robbertsei* as a species that could have evolved from *V. karroo* and *V. gerrardii* genes. The locations where these different forms (previously) recognised within the *V. karroo* complex occur are shown in **Figure 4**.

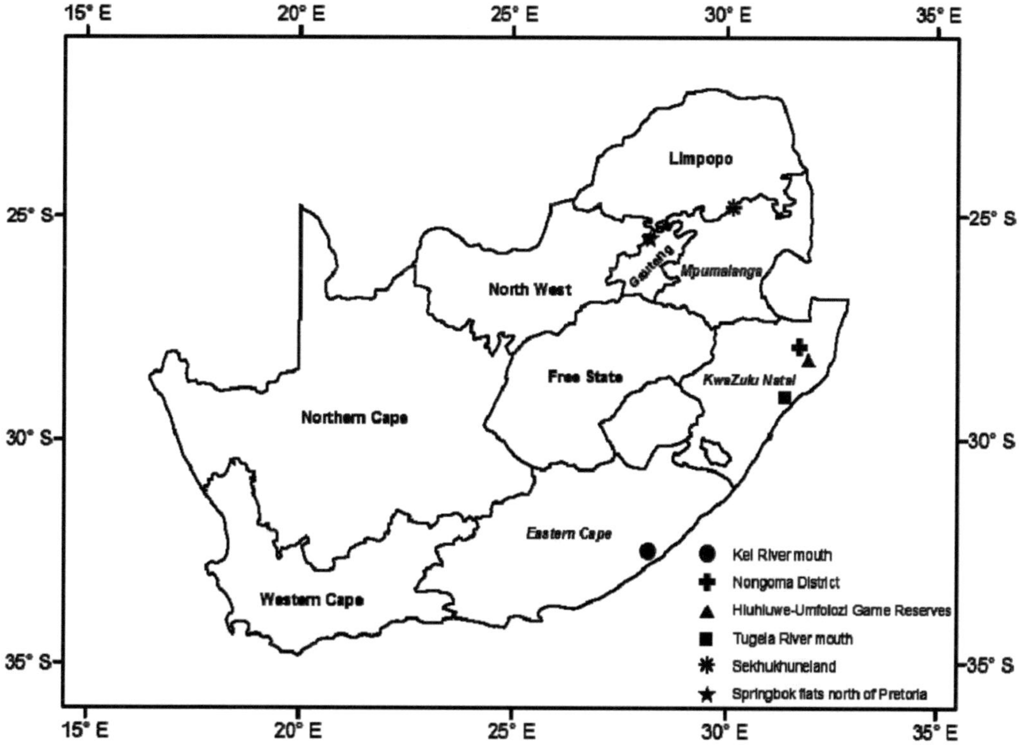

Figure 4. Map indicating occurrences of the various morphological variations (described in **Table 2**) within the *Vachellia karroo* complex. Note: The three provinces labelled in italics indicate occurrence of variation 1.

2. Ecological significance

This is an ecologically and socio-economically important species described by many as a multi-purpose tree and an asset to any farm [18].

2.1. Value as fodder and food supplement

Vachellia karroo attracts many insects and therefore birds, and its flowers also form an important food supplement for animals [1]. The flowers have significant amounts of pollen and are rich in protein and are thus eaten by birds such as Grey Go-Away birds (also known as Grey Louries) and monkeys [5]. In addition, the larvae of several butterfly species feed on the pods and flowers [5, 16]. Its flowers also provide nectar for bees and are important for the production of honey [1]. *V. karroo* trees are important for bee farming as they indirectly result in the production of a pleasantly flavoured honey [6].

Furthermore, parts of *V. karroo* are used as food for humans, as an example, seeds are roasted and used as coffee substitute [19]. *Vachellia* species can produce large amounts of seeds which are known to have been eaten by pastoral people when the need arose, and indirect food sources include the edible cerambycidae wood borer larvae found in the dead wood of *V. robusta* [1].

Despite its thorniness, *V. karroo* is a good fodder tree and forms an important part of the diet of a wide range of herbivore species [2]. It is palatable [20] and consumed by both domestic and wild species [21]. Its foliage is highly favoured by stock and game, so are its seeds and dehiscent pods, which are rich in protein [18]. The pods and seeds also play an important role as feed supplements during the dry season [22] as they are at times collected by farmers to feed their livestock [1]. *V. karroo* has been shown to be an important part of the giraffe diet [23], and it has also been observed that goats select *V. karroo* in preference to grass, but less so, when the amount of available browse available is limited [21]. The foliage, pods, and flowers of *V. karroo* are free of hydrocyanic poisoning, a self-protection mechanism used by many trees [6], relating to the toxic substance known as hydrocyanic acid, prussic acid, or cyanide. Some *Vachellia* species pose the danger of such poisoning to animals. These include *V. erioloba*, whose pods and young leaves contain prussic acid, as well as the wilted leaves of *V. sieberiana* [1].

2.2. Bush encroachment

Vachellias are, on the whole, easy to grow and often become an aggressive invader of valuable farming land and grazing areas, a phenomenon that is usually referred to as bush encroachment (**Figure 5**). Bush encroachment has become a serious ecological and farming problem that has affected many grazing areas in grassland and savanna areas of southern Africa. It is

Figure 5. *Vachellia karroo* encroachment of a grassland community, Free State Province, South Africa (photo: M. Dingaan).

a transition from grassy to increasingly shrubby ecosystems [24], whereby trees and shrubs invade into open grassland or thicken up in already wooded areas [25]. *V. karroo*, in particular, has become a serious invader into the grasslands of the Eastern Cape, the Free State area, and the North-West Province [26, 27].

The most detrimental effect of *V. karroo* encroachment (and other woody species) to farming is that it depresses the production of grasses, mainly due to tree-grass competition for soil moisture [28]. Bush encroachment thus drastically reduces the carrying capacity of grazing areas because browse is generally a poor substitute for grass, especially in sheep/cattle areas [29]. For example, in some parts of the Molopo area, grass production was thought to have already decreased by over 80% due to bush encroachment, and this has subsequently affected the economic viability of many farms [30].

2.2.1. *Factors promoting bush encroachment*

Vachellia species regenerate vegetatively and from seed, but regeneration from seed is most dominant [31]. The encroaching species of *Vachellia* are spread by seed, which in many of these species has impermeable seed coat resulting in a high percentage of dormancy [32]. According to O'Connor [33], the encroachment of woody species requires successful seed dispersal, germination, and seedling establishment. The two most vulnerable phases in the regeneration of *Vachellia* are during seed germination and seedling establishment; these phases are characterised by high mortality rates that influence the populations of *Vachellias* [31]. Seedling establishment can be influenced by competition from established surrounding vegetation, as well as moisture availability and irradiance [33]. Du Toit [26] has shown that *V. karroo* seedlings require high irradiance levels for optimal growth, although they may still survive under certain levels of low irradiance.

According to Trollope [25], the plausible reasons why bush encroachment was not a serious problem before the advent of commercial livestock production could have included the control of bush by fire, mechanical damage brought about by wild browsers and climatic factors. On the other hand, the factors promoting encroachment in the modern era are complex, with the most predominant being the introduction of domestic livestock and subsequent overgrazing and the elimination of veld burning [24, 33, 34]. The successful and prolific nature in which *V. karroo* has been able to encroach onto grasslands is largely due to the fact that the species is an adaptable pioneer with an ability to establish itself without shade, shelter or protection from grass fires. It is fast growing, tolerant of defoliation by herbivory and is resistant to fire and frost [18]. Its seeds do not only have a great tolerance to high temperatures produced during burning but may actually be stimulated to germinate by fire [35].

2.2.2. *Combating bush encroachment*

Clearing of woody species has been found to greatly increase grass and subsequently animal production [28, 30]. Mechanical, chemical and biological methods are employed in trying to control the spread of bush. Chemicals such as Tordon 225 and tebuthiuron have been successfully used, but the use of Tordon 225 is restricted by certain physiological and environmental conditions [30]. Concerns about tebuthiuron on the other hand pertain to the accumulation and persistence of the chemical in the soil thus posing potential threats to non-target species [36, 37]. Biological methods sometimes employed include the controlled use of herbivores (especially goats) and fire. Du Toit [29] observed in a study in the Eastern Cape that in comparison to continuous/rotational sheep grazing of a *V. karroo* stand, there was a higher mortality of trees and more efficient control of seedling regrowth under continuous goat grazing than rotational grazing. Goat grazing resulted in a marked improvement in the cover, composition and vigour of the grass sward [29].

Fire has also been extensively used in combating bush encroachment in savanna because it is known to maintain a balance of grass to trees and shrubs in the savanna areas [25]. Trollope [25] has observed that fire generally has different roles in controlling bush encroachment in the moist and arid savannas. In the moist savanna regions (>600 mm p.a.), bush encroachment may be controlled with fire alone because there is adequate grass material under grazing conditions to support frequent enough fires to burn down and control the bush. This is unlikely in the arid savanna regions (<600 mm p.a), which constitute the major portion of the South African savannas, because the rainfall is too low and erratic to support frequent enough fires under grazing conditions to prevent the regeneration of bush [25]. In grassland, Du Toit [34] made observations that the application of fire to combat *V. karroo* intrusion in the Eastern Cape sweetveld was not a practical approach. While fire was found to retard *V. karroo* seedling development, it could however not prevent the seedling establishment.

All in all, eradication of *V. karroo* is difficult once the thorn has invaded an area where it was previously absent, since a seed bank which did not previously exist is established. *V. karroo* trees can produce large amounts of seeds annually, and these have a high longevity. As a result, destruction of a stand of *V. karroo* is often times still followed by seedling establishment and considerable regeneration [34].

2.3. Role in soil fertility

The effect of *V. karroo,* and other tree species, on herbaceous species (and grasses) may not always be negative, and there is evidence that trees may actually have a beneficial effect on neighbouring plants. For example, increased herbaceous layer productivity has been reported under tree canopies, due to favourable conditions such as improved soil water status and soil fertility [38, 39]. Likewise, *V. karroo* has various favourable influences on herbaceous production. First, *V. karroo* is a leguminous tree known to form root nodules [40], which are swellings on the root that contain nitrogen-fixing microorganisms (bacteria) known as *Rhizobium. Rhizobium* possesses the enzyme systems (including nitrogenase enzyme complex) that convert atmospheric nitrogen to nitrogen compounds useful to plants [41, 42]. Legumes like *V. karroo* then use the compounds to construct amino acids and protein [41, 43]. This ability of *V. karroo* to fix nitrogen is beneficial to other plants as well, mainly because the nitrogen content in the soil increases, and soil fertility is thus enhanced under these trees. In addition, *V. karroo* is able to use water and nutrients from deep underground because it has a long taproot, and this again leads to grasses and other plants thriving in its shade [6]. The ability of *V. karroo* to use water from deep underground means that it can grow in arid and otherwise inhospitable environments, as long as there is an assured supply of underground water [27]. It hence also acts as an indicator of surface and underground water, especially in arid land [2, 5]. The tree is further considered an indicator of sweet veld, which is highly valued for good grazing and fertile soils [5, 6]. This is due to the beneficial effects such as provision of shade, improved soil fertility, and water availability, which lead to the development of palatable and nutritious grasses under the *V. karroo* trees [18].

Several studies have been conducted on the positive effect of woody plants on grasses. In southern African savannas, *Panicum maximum* is well known to be associated with tree canopies, especially those of several *Vachellia* species. *P. maximum* is one of the most important fodder grass species in many savanna areas, mainly because it is highly palatable to cattle and other grazers, and it also has a high production potential [44]. The grass is strongly associated with tree canopy cover; it is common under trees but seldom occurs in the open [45]. Smit and Swart [46] suggest that such grass-tree associations, which exist in many semi-arid savanna areas, warrant that bush control measures should not simply imply a complete removal of woody plants but rather tree thinning with a view to reducing negative competition effects. This kind of approach can ensure that the important forage contribution by *P. maximum* is maintained.

This association is likely due to enhanced supply of nutrients such as nitrogen and phosphorus under tree canopies and suitable germination conditions for *P. maximum* seeds due to the relative abundance of litter and low temperatures under tree canopies [1]. A study investigating the relation between tree height of *V. karroo* and *V. tortilis* and the associated occurrence of *P. maximum* in the Sourish Mixed Bushveld [27] of Limpopo Province indicated that *P. maximum* mainly occurs under larger trees, but the grass attained pure stands under smaller *V. tortilis* trees of >2.0 m height [44]. In the False Thornveld of the Eastern Cape [27], Stuart-Hill et al. [47] proposed that the net effect of the favourable or unfavourable influences of *V. karroo* on grass production is dependent on tree density. It was observed that in situations where there were a few *V. karroo* trees, grass production was greater than where there were no trees but declined as tree density increased beyond a critical level.

3. Socio-economic significance and uses

V. karroo is of considerable socio-economic value as almost all of its parts, including bark, pods, seeds, leaves and thorns, are extremely useful to both humans and animals.

3.1. Domestic uses

Vachellia karroo is one of the most preferred species for fuelwood [48] because the wood has excellent fuel properties. It burns clean with little smoke and is valued for its sustained high temperature [18] and thus produces high-quality fuelwood for many rural communities which still rely on wood for cooking and heating. The wood is also used as rough construction material for building traditional huts and fences in many rural communities [1, 18]. The thorns are used as sewing needles, pegs or pins, while its branches are used in farms to make fencing kraals for livestock, to protect them from predators [1, 2]. The bark, leaves, gum and other parts are used medicinally in many ways. An infusion of the bark is used to cure diarrhoea and dysentery, while the dried and powdered form of its gum is used for eye treatments [2]. A boiled liquid from the bark is sometimes used to treat cattle which have tulp poisoning caused by *Moraea* (*Homeria*) species, which are bulbous plants poisonous to cattle [6]. Other *Vachellia* species are known to have medicinal properties as well. For example, the bark of *V. erioloba* is used to treat headaches and that of *V. xanthophloea* is used for fevers and eye complaints [1].

3.2. Commercial value

In addition to all the domestic uses of *V. karroo*, various commercial products are also obtained from the tree, of which gum is one of the most important (**Figure 6**). In fact, *V. karroo* gets its common name "sweet thorn" from this gum which comes out from wounds in the bark [6]. It is a pleasant tasting gum that is eaten by people and animals and has also been used for confectionary and adhesives [2, 16]. This gum is similar to gum arabic, which is widely used for thickening many convenience foods, pharmaceuticals and cosmetics [1].

The wood of *V. karroo* is hard and tough, making it suitable for making furniture, poles, and fence-posts [2]. It is also used to make wooden carvings (ornaments), which are very popular ornaments in the tourism industry [1]. The bark is used to make strong ropes and mats [2]. This bark and that of several other *Vachellia* species, notably *V. nilotica* (bark and pods)

Figure 6. *Vachellia karroo* tree exuding gum (photo: M. Dingaan).

contains tannin [1], which is widely used in the tanning of leather, giving it a reddish colour [5]. Tannins are plant polyphenolic compounds (secondary metabolites) that act as a defence mechanism in plants against pathogens and herbivores [49–51] and hostile environmental conditions [52, 53]. Most of the commercially extracted tannin in South Africa comes from Black Wattle (*Acacia mearnsii*), an introduced Australian species which can yield 36–44% tannin from the bark [1].

4. Occurrence and distribution of *V. karroo* in South Africa

A TWINSPAN classification of historical data comprising 1553 relevés and 2006 species, compiled from all areas of South Africa where *V. karroo* is known to occur, was conducted and produced five main vegetation types, namely savanna, grassland, riparian thickets, wetland and Nama-Karoo communities. The riparian thickets and Nama-Karoo communities will only be mentioned briefly in this section, because they form the core of the *Acacia* (*Vachellia*) *karroo* Class suggested and described in more detail by Dingaan [54].

4.1. Savanna communities

Savannas are one of the main biomes in the world and are the dominant vegetation in Africa and southern Africa [55], especially in Botswana, Namibia and Zimbabwe [56]. In South Africa, the Savanna forms the largest biome and occupies over one-third (33.49%) the country's area [57]. It is well developed in Northern Cape, North-West, and Limpopo Provinces; it is also found in parts of Mpumalanga, KwaZulu-Natal, and Eastern Cape Provinces and has isolated occurrences in Gauteng and Free State Provinces. The factors delimiting the biome are complex and can be an interplay of altitude, climate, soils, herbivory, and fire [39, 56, 58]. The biome mostly occurs at altitude ranging from sea level to 2000 m; rainfall is seasonal with wet summers and dry winters and varies from 200 to 1000 mm per year (**Figure 7**) frost may occur from 0 to 120 days per year, with frost free days in low-lying areas and longest frost periods in high-altitude areas [56, 58]. Approximately 8.5% of the biome is conserved in South Africa [57], a fairly good proportion compared to the other biomes. There are several conservation areas in the biome, which include the Kruger National Park. Savanna areas have not been adversely impacted by urbanisation, which could have been hindered by the hot, moist climate and diseases such as malaria [56, 57].

The Savanna biome in South Africa is described by Low and Rebelo [56] and Scholes [55] as vegetation characterised by a grassy ground layer 0.5–2 m tall and a distinct upper layer of woody plants 2–10 m tall (**Figure 8**). It may be delineated according to the height and degree of canopy cover of the tree layer as follows: shrubland, woodland, or bushveld depending on whether the upper layer is near the ground, dense or in the intermediate stages, respectively [55, 56]. Savanna vegetation may be broadly divided into fine-leaved savannas found in nutrient-rich and arid environments and broad-leaved savannas in nutrient-poor and moister environments [55, 58]. Broad-leaved species such as *Terminalia sericea, Burkea africana,* various *Combretum* species, *Pterocarpus rotundifolius* and several others dominate the higher rainfall

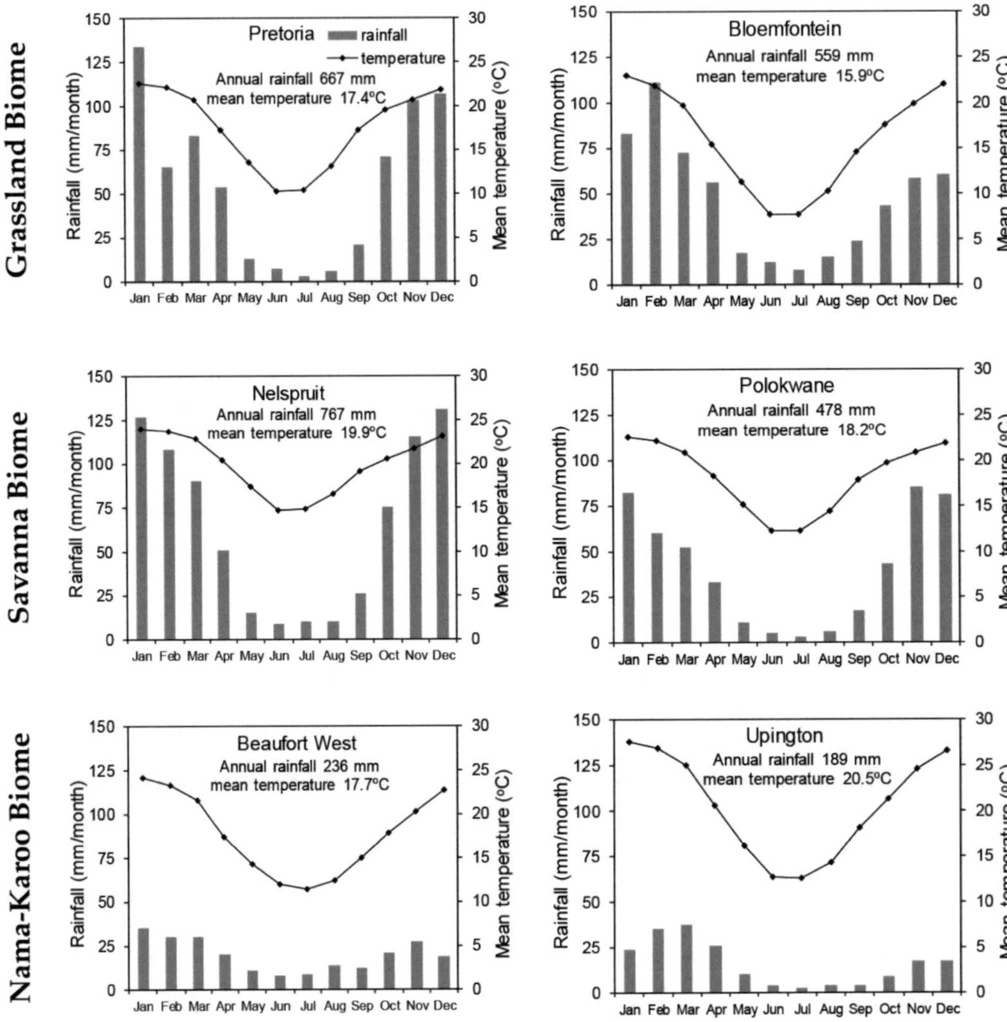

Figure 7. Rainfall and temperature for selected locations in the three representative biomes of South Africa [54].

areas, while the more arid savanna is dominated by microphyllous species where numerous *Vachellia* (and *Senegalia*) species dominate the tree component, but *Colophospermum mopane* is the broad-leaved exception [55].

The areas of the Savanna biome where *V. karroo* occurs are mainly in the Limpopo and North-West Provinces (**Figure 9**), and vast communities of *V. karroo* also occur in the Kalahari region in the Northern Cape Province. It can also be found in parts of the Eastern Cape and Western Cape Provinces. *V. karroo* communities can be encountered on predominantly sandy soils on bottomlands, footslopes, and mountain slopes. They can also be found as riparian thicket on clayey soils along stream and riverbanks.

Communities that make up this vegetation type are listed in **Table 3**, with the two most prominent as follows: The first is the *Acacia karroo–Panicum maximum* Open Woodland [62]

Figure 8. Savanna near Kimberley (top) and at Mokala National Park (middle, bottom), Northern Cape, South Africa (photos: M. Dingaan).

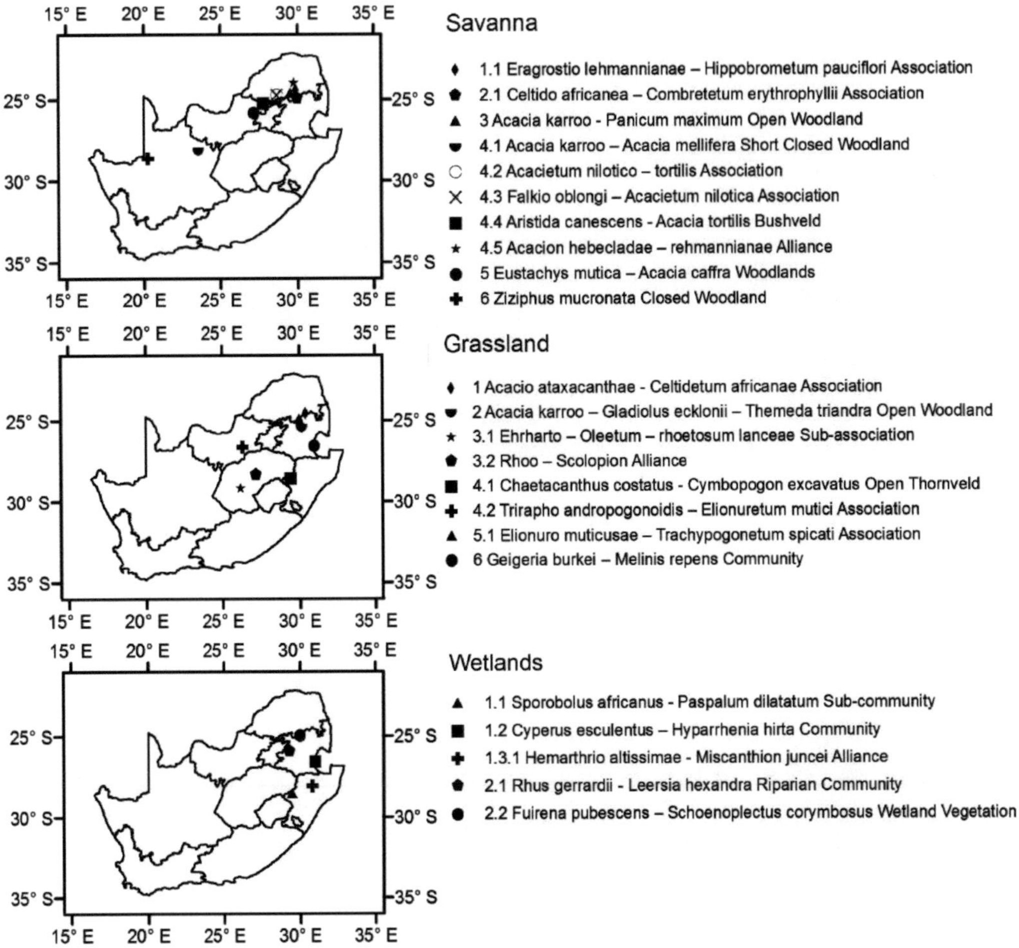

Figure 9. Distribution of selected *Vachellia karroo* communities.

found mainly along the banks of Ngwaritzi and Olifants Rivers south of Polokwane in the Limpopo Province. The second is the *Panico maximi–Acacietea tortilis* Class, described by Winterbach et al. [59] as microphyllous thorny bushveld that is associated with dark, clayey soils in low-lying areas. Other communities in the savanna where *V. karroo* occurs include the *Kirkia wilmsii–Terminalia prunioides* Closed Mountain Bushveld described by Siebert et al. [61]. This vegetation occurs within the Sekhukhuneland Centre of Plant Endemism (SCPE), which stretches from the Limpopo Province into the Mpumalanga Province and includes towns such as Roossenekal, Steelpoort and Sekhukhune. The vegetation is predominantly restricted to the warm slopes and valleys of undulating hills and mountains.

4.2. Grassland communities

The South African Grassland biome is part of the global temperate grassland biome [69]. It is the third largest biome in the country and covers 25.71% of South Africa [57] The biome is

Community	Location and habitat	Refs.
1. *Englerophyto magalismontani–Acacietea caffrae* Class	Found on slopes of Waterberg, Magaliesberg, Witwatersrand, Suikerbosrand mountains (Limpopo) Mainly on sandy soils	[59]
1.1. *Eragrostio lehmannianae–Hippobrometum pauciflori* Association	Sekhukhuneland (SCPE) (Limpopo/Mpumalanga) Occurs as scattered thickets in large river valleys, dongas or eroded areas Sandy soils, mainly Hutton form	[60]
2. *Kirkia wilmsi–Terminalia prunioides* Closed Mountain Bushveld	Sekhukhuneland (SCPE) (Limpopo/Mpumalanga) Restricted to warm slopes and valleys of undulating hills and mountains Soils are generally clayey Surface rocks are predominant	[61]
2.1. *Celtido africanea–Combretetum erythrophyllii* Association	Closed riparian thicket to forest found in valleys along large rivers such as the Steelpoort and Olifants Red loam Oakleaf soils	[60]
3. *Acacia karroo–Panicum maximum* Open Woodland	Polokwane (formerly Pietersburg), Limpopo	[62]
3.1. *Sporobolus ioclados–Gymnosporia buxifolia* Short Closed Woodland	Occurs along the banks, floodplains of the Ngwaritzi and Olifants Rivers Deep clayey, poorly drained soils Oakleaf, Dundee, and Valsrivier forms	[62]
3.2. *Commiphora africana–Digitaria velutina* Low Open Woodland	Found on plains, along the upper banks of the Ngwaritzi River and along the ridges Shallow, well-drained sandy soils Hutton and Glenrosa forms	[62]
4. *Panico maximi–Acacietea tortilis* Class	Waterberg, Soutpansberg and Pietersburg Plateaus (Limpopo) Dark, clayey soils	[59]
4.1. *Acacia karroo–Acacia mellifera* Short Closed Woodland	Rhino Ranch, Lephalale (formerly Ellisras), Limpopo Clayey soils of the Arcadia form	[63]
4.2. *Acacietum nilotico–tortilis* Association	Nylsvley Nature Reserve, Limpopo Found on flat bottomlands, typically on calcareous alluvium Soils mainly Oakleaf, Valsrivier, and Arcadia forms	[64]
4.3. *Falkio oblongi–Acacietum nilotica* Association	Nylsvley Nature Reserve, Limpopo Found on bottomland Vertic soils of the Arcadia form	[64]
4.4. *Aristida canescens–Acacia tortilis* Bushveld	Borakalalo Nature Reserve (BNR), North-West	[64]
4.4.1. *Acacia erubescens–Acacia luederitzii–Plectranthus madagascariensis* Thornveld	Dry to moderately dry habitat Sandy, sometimes clayey soils	[65]
4.4.2. *Perotis patens–Terminalia sericea* Woodland	Found on southern section of BNR Sandy soils	[66]

Community	Location and habitat	Refs.
4.4.3. *Ziziphus mucronata–Acacia karroo* Woodland Note: Although this community is similar to the communities of the *Vachellia karroo* Class suggested by Dingaan [54] in that it is *Vachellia karroo*–dominated riparian vegetation, it shows more affinity towards communities of the *Panico maximi–Acacietea tortilis* and is hence correctly included by Winterbach [64] in this class	Situated on the banks of the Moretele River and tributaries (BNR) Loamy to clayey soils, sandy soils in some tributaries	[65]
4.5. *Acacion hebecladae–rehmannianae* Alliance	Vicinity of Turfloop Dam, Limpopo Mainly occurs along streams and adjacent areas Clayey soils	[64]
5. *Eustachys mutica–Acacia caffra* Woodlands	Kgaswane Mountain Reserve (formerly Rustenburg Nature Reserve), North-West Found on slopes of the Magaliesberg Also on flat surfaces with clay-loam soils	[67]
6. *Ziziphus mucronata* Closed Woodland	Augrabies Falls National Park, Northern Cape Associated with drainage lines, floodplains and islands of the Orange River Dominant soil forms are Dundee and Oakleaf	[68]

Table 3. Classification and habitat features of savanna communities.

found mainly on the high central plateau (Highveld) comprising the Free State and Gauteng Provinces and is also found in parts of Mpumalanga Province and the inland areas of KwaZulu-Natal and Eastern Cape Provinces. Most of the large urban areas are concentrated in the biome, and consequently, the grassland biome has the greatest urban population density in South Africa [57]. The urban expansion, coupled with conversion of natural grassland to cultivated land, has resulted in a huge decline in biodiversity in this biome [70]. Most of the grassland is converted for the production of crops such as maize, wheat, sorghum and sunflower. Compared to the savanna, conservation of grasslands is relatively low with only 1.12% of the biome conserved [57].

The distribution of the biome is determined by an interplay of climate, topography, fire and grazing [71]. The overall extent of the biome is mainly determined by climate, especially the amount of summer rainfall and minimum winter temperatures [69]. The grass dominance is maintained by frosts, fire and grazing, which also prevent the establishment of trees [56]. However, the role of fire in maintaining grassland is greater in humid (>650 mm of annual rainfall) than semi-arid regions (<650 mm of annual rainfall) [69, 71]. The biome is limited to altitudes varying from near sea level to 2850 m above sea level; the winters are cold, dry with frequent occurrences of frost; rainfall varies spatially from 400 to 2500 mm per annum and occurs mainly during the summer season [57, 69]. The topography is mainly flat to slightly undulating and may include mountainous regions [69].

http://dx.doi.org/10.5772/intechopen.70456

The biome comprises grasslands that are dominated by a single layer of grasses (**Figure 10**), with forbs forming an important but usually not dominant component. The dominant grasses in the biome are of the genera *Andropogon, Cymbopogon, Diheteropogon, Heteropogon, Hyparrhenia, Monocymbium, Schizachyrium, Themeda, Trachypogon* and *Tristachya* [69]. Trees are generally absent, except in a few localised habitats. The woody component is usually limited to higher

Figure 10. Grasslands near Bloemfontein (top), Bethlehem (middle) and Winburg (bottom), Free State Province, South Africa (photos: M. Dingaan).

moisture areas such as hills, gullies, valley slopes and is also found on azonal alluvial soils. The woody species often found in grassland are *V. karroo, V. sieberiana,* species of *Protea, Cussonia, Diospyros, Gymnosporia* and many more. Some of these trees and shrubs can tolerate frequent fires by being serotinous and through their ability to resprout after fires [69]. The Grassland biome can be divided into two classes (sweet and sour grasslands), based on moisture availability and palatability to livestock. Sweet grasslands (locally known as sweetveld) are dry grasslands that occur on base-rich soils at lower altitudes and remain palatable and nutritious throughout the year. Sour grasslands (sourveld) are moist grasslands generally found on leached soils at higher altitudes, which are palatable only in spring and summer [69, 72].

Vachellia karroo occurs throughout the biome and often encroaches on degraded grasslands. It is found on plains where soils are sufficiently deep, as well as in sheltered sites on the slopes, where habitat conditions are relatively moist. Communities where it is found are listed in **Table 4** and the two most prominent are as follows: The first is the *Themeda*

Community	Location and habitat	Refs.
1. *Acacio ataxacanthae–Celtidetum africanae* Association	North-eastern Mpumalanga and south eastern Limpopo Mountain sourveld on dry dolomitic regions Rock outcrops near or on the bottom of valleys Some protected areas on valley sides	[73]
2. *Acacia karroo–Gladiolus ecklonii–Themeda triandra* Open Woodland	Northern Mpumalanga Belfast-Lydenburg-Dullstroom area Plains and slopes Diverse soil types and forms	[74]
3. *Rhoetea erosae* Class Originally described by Werger [75] In present classification, it represents the shrub communities of southern and eastern Free State as described by Du Preez and Bredenkamp [76]	(i) Shrub communities occurring along the Upper Orange River Valley (ii) Shrub communities typical of the talus slopes of mountains, dolerite hills and ridges Also includes grassy shrubland communities on low dolerite outcrops	[75] [77]
3.1. *Ehrharto–Oleetum–rhoetosum lanceae* Sub-association Part of the *Chrysocomo–Selagenea albidae* sub-class [77]	Bloemfontein, Free State Province Relatively moist habitats found in gorges and drainage lines on the slopes of dolerite hills	[77]
3.2. *Rhoo–Scolopion* Alliance Synonym: *Grewio–Isoglossion grantii* Alliance Du Preez [77] Part of the *Rhoo–Rhoicissenea tridentatae* sub-class proposed by Du Preez [77]	Willem Pretorius Nature Reserve, Winburg-Ventersburg area, Free State Province Shrubland occupying plateaus and steep slopes of dolerite hills, rocky outcrops of the Beaufort Formation	[78]
4. *Themeda triandra–Eragrostis plana* Class	Moist grasslands of the plains High altitudes and high rainfall	[76]
4.1. *Chaetacanthus costatus–Cymbopogon excavatus* Open Thornveld Described by Robbeson [79] as Open Thornveld, a variation of Acocks' [27] Southern Tall Grassveld	North-western KwaZulu-Natal Includes the towns Estcourt, Colenso and Ladysmith, as well as Bergville and Winterton Plains adjacent to the footslopes of the Drakensberg Soils mostly shallow, sandy or sandy loam	[79]
4.1.1. *Hermannia depressa–Anthospermum rigidum* Sub-community	Slopes and footslopes of rocky hills Deep sandy soils	[79]

Community	Location and habitat	Refs.
4.1.2. *Hyparrhenia hirta–Themeda triandra* Grassland Community	Plains close to rocky hills Sandy and clayey soils	[79]
4.1.3. *Scabiosa columbaria–Aster peglerae* sub-variation	Open grassland on slopes of rocky hills Shallow soils of the Mispah form	[79]
4.2. *Trirapho andropogonoidis–Elionuretum mutici* Association	Ottosdal-Delareyville-Lichtenburg area, North-West Province High altitude grassland on midslopes Well-drained, sandy soils Mainly Hutton, Avalon and Mispah soil forms	[80]
i) *Helichrysum rugulosum–Conyza podocephala* Grassland	Pretoria and Heidelberg area, Gauteng Province Witbank, Mpumalanga Province Moist, deep soils on the undulating and flat plains Dominant soil forms are Glenrosa, Clovelly, and Hutton	[81]
5. *Tristachya leucothrix–Trachypogon spicatus* Class Synonym: *Harpochloo–Tristachyetea leucothrichis* Class Du Preez [77]	Korannaberg, Clocolan, Ficksburg, Bethlehem, Golden Gate, Platberg mountain near Harrismith, Free State Province Moist, high altitude mountain slopes and plateaus Sandy soils	[76]
6. *Geigeria burkei–Melinis repens* Community	Southeastern Mpumalanga Belfast–Barberton–Piet Retief–Wakkerstroom area Mountains and plains Associated with sandy loam soils Glenrosa the dominant soil form	[83]
6.1. *Acacia nilotica–Aristida congesta* Community	Found on strongly undulating plains Sandy to sandy loam soils Glenrosa soil form dominant	[83]
6.1.1. *Perotis patens–Hyperthelia dissoluta* Pure Short Closed Grassland Community	Found on slopes characterised by sandy to sandy loam soils Hutton and Glenrosa forms	[83]
6.1.2. *Pavetta edentula–Pellaea calomelanos* Low/Short Thicket Community	Occurs along the crest of strongly undulating plains Soils generally sandy to sandy loam, and rocky Mainly Glenrosa form	[83]
6.2. *Dombeya rotundifolia–Heteropogon contortus* Low/Short Thicket Community	Mountain vegetation associated with north facing slopes Sandy clay loam to sandy clay Glenrosa soils	-

Table 4. Classification and habitat features of grassland communities.

triandra–Eragrostis plana Class proposed and described by Du Preez and Bredenkamp [76] as moist grasslands of the plains at relatively high altitudes and high rainfall. The second is the *Geigeria burkei–Melinis repens* community; the individual communities that represent this vegetation type were identified and described by De Frey [83], but they are classified together under one major community for the first time in the present classification. This is vegetation of the mountains and plains of southeastern Mpumalanga, specifically the area comprising the towns of Belfast, Barberton, Piet Retief, and Wakkerstroom. It is associated with sandy loam soils, with Glenrosa as the dominant soil form.

4.3. Riparian thickets

The riparian thickets dominated by *V. karroo* are mainly associated with deep, clayey alluvial deposits that occur along stream and river banks (**Figure 11**) and occasionally on

Figure 11. Riparian vegetation along the Modder River near Glen (top, middle) and Sand River near Ventersburg (bottom), Free State Province, South Africa (photos: M. Dingaan).

the river beds. The thickets also extend to the floodplains and bottomlands adjacent to the watercourses and also on gradual footslopes of hills and ridges. This vegetation type forms the core of the *Acacia* (*Vachellia*) *karroo* Class suggested and described in more detail by Dingaan [54].

4.4. Wetland communities

Wetland communities in which *V. karroo* is usually encountered are found in KwaZulu-Natal, Mpumalanga and Limpopo Province. Although these communities occur in both the Savanna and Grassland Biomes, we regard them as a distinct vegetation type because of their unique species composition. This vegetation type differs from the riparian thickets, which are mainly associated with clayey soils along rivers and streams. The wetland communities described here are generally associated with moist sandy soils and are dominated by grasses and forbs. *V. karroo* in these communities is the only notable woody species (**Figure 12**) but is not as prominent as in the riparian thickets.

Some of the major communities recognised within this vegetation type are as follows: The first is the *Hemarthria altissima* Class described by Du Preez and Bredenkamp [76] for the southern and eastern Free State. It represents vegetation of moist soils on marshes, streambanks, riverbanks, dam edges, and vleis (shallow, seasonal wetlands). Although *V. karroo* is not present in communities described by Du Preez and Bredenkamp [76], it can be encountered in other wetland communities regarded as part of this class, namely those of the central-northern KwaZulu-Natal described by Eckhardt et al. [82]. The other distinct community is the *Fuirena pubescens–Schoenoplectus corymbosus* wetland vegetation described by Siebert et al. [61]. This wetland vegetation is found throughout the Sekhukhune Centre of Plant Endemism in Limpopo and Mpumalanga Provinces. It occurs on stream banks in valleys, in seepage areas on mountain slopes and also in wetlands on the mountain plateaus. It is associated with wet, vertic black clay soils. The main distinction between this vegetation and other *Vachellia karroo*-dominated riparian thickets is the absence of woody species such as *Ziziphus mucronata, Diospyros lycioides,* and *Rhus pyroides,* which are the usual companions of *Vachellia karroo* along the riverbanks. The wetland communities where *V. karroo* occurs are listed in **Table 5**.

4.5. Nama-Karoo communities

The Nama-Karoo biome is the second-largest biome in South Africa, covering 28.35% of the country [57]. It occurs on the western half of South Africa, at altitudes ranging from 500 to 2000 m but most of the biome falls between 1000 and 1400 m [56, 57]. This is an arid biome, characterised by unreliable summer rain that varies between 100 and 520 mm per year [85]. The topography resembles extensive, flat to undulating plains dotted with hills and occasional mountains [57]. The dominant vegetation is a grassy, dwarf shrubland (**Figure 13**), comprising a mix of low shrubs, grasses, succulents, geophytes and annual herbs [56, 85]. The annuals on average comprise the highest number of species in the biome [86].

V. karroo in this karroid vegetation is found in southern Free State and some areas in the Eastern Cape. The vegetation is found in varied habitats, ranging from gentle slopes

Figure 12. Wetlands near Verkeerdevlei (top), Winburg (middle) and Ventersburg (bottom), Free State Province, South Africa (photos: M. Dingaan).

http://dx.doi.org/10.5772/intechopen.70456

Community	Location and habitat	Refs.
1. *Hyparrhenia dregeana–Eragrostis plana* Wetland	KwaZulu-Natal and Mpumalanga Rivers and streams	-
1.1. *Sporobolus africanus–Paspalum dilatatum* Sub-community	North-western KwaZulu-Natal Estcourt-Colenso-Ladysmith area Occurs in riverbeds Deep, sandy loam to clayey soils	[79]
1.2. *Cyperus esculentus–Hyparrhenia hirta* Community	Mountain wetland of the Belfast-Barberton-Piet Retief-Wakkerstroom area, Mpumalanga Occurs in valley bottoms Sandy clay loam to sandy clay soils Katspruit dominant soil form	[83]
1.3. *Hemarthria altissima* Class	Southern and eastern Free State Restricted to marshes and stream banks on the plateaux of the Korannaberg Also found in the Willem Pretorius Game Reserve on riverbanks, dam edges and wetlands with permanent water	[77] [78]
1.3.1. *Hemarthrio altissimae–Miscanthion juncei* Alliance Originally described by Eckhardt et al. [82] as an alliance of the *Agrostis lachnantha–Eragrostis plana* Wetlands	Central-northern KwaZulu-Natal Helpmekaar-Utrecht-Louwsburg area Rivers and streams Alluvial sandy soil Predominantly Dundee form	[82]
2. *Conyza scabrida–Gomphostigma virgatum* Wetland	SCPE (Limpopo/Mpumalanga) and Witbank Nature Reserve(Mpumalanga) Occurs along rivers and streams, and in the rocky streambeds	–
2.1. *Rhus gerrardii–Leersia hexandra* Riparian Community	Witbank Nature Reserve, Mpumalanga Occurs along the banks of the Olifants River and in the rocky streambed Soils generally sandy, rocky and shallow Predominantly Glenrosa and Mispah soil forms.	[84]
2.2. *Fuirena pubescens–Schoenoplectus corymbosus* Wetland vegetation	Throughout SCPE, Limpopo/Mpumalanga Found on stream banks in valleys, in seepage areas on mountain slopes, and also in wetlands on the mountain plateaus Associated with wet, vertic black clay soils	[61]

Table 5. Classification and habitat features of wetland communities.

and plateaus in south-western Free State to rocky habitats on hot and dry slopes in the Eastern Cape. The presence of *V. karroo* in these karroid veld types can be ascribed to bush encroachment occurring as a result of overgrazing [27]. The plant communities where *V. karroo* occurs are part of the *Acacia (Vachellia) karroo* Class proposed and described in detail by Dingaan [54].

Figure 13. Nama-Karoo at Augrabies National Park (top, middle) and in Hopetown (bottom), Northern Cape Province, South Africa (photos: M. Dingaan).

http://dx.doi.org/10.5772/intechopen.70456

Author details

Mamokete Dingaan[1]* and Pieter J. du Preez[2]

*Address all correspondence to: dingam@unisa.ac.za

1 Department of Life and Consumer Sciences, University of South Africa, Florida, South Africa

2 Applied Behavioural Ecology and Ecosystem Research Unit, University of South Africa, Florida, South Africa

References

[1] Smit N. Guide to the Acacias of South Africa. Pretoria: Briza Publications; 1999

[2] Palmer E, Pitman N. Trees of Southern Africa. Vol. 2. Cape Town: A.A. Balkema; 1972

[3] Carr JD. The South African Acacias. Johannesburg: Conservation Press; 1976

[4] Davidson L, Jeppe B. Acacias: A Field Guide to the Acacias of Southern Africa. Johannesburg: Centaur Publishers; 1981

[5] Coates PM. Keith Coates Palgrave Trees of Southern Africa. 3rd ed. Cape Town: Struik Publishers; 2002

[6] Aubrey A, Reynolds Y. *Acacia karroo* Hayne [Internet]. 2002. Available from: http://www.plantzafrica.com/plantab/acaciakar.htm [Accessed: January 24, 2017]

[7] Maslin BR, Seigler DS, Ebinger J. New combinations in *Senegalia* and *Vachellia* (Leguminosae: Mimosoideae) for Southeast Asia and China. Blumea. 2013;**58**:39-44

[8] Maslin BR, Miller JT, Seigler DS. Overview of the generic status of *Acacia* (Leguminosae: Mimosoideae). Australian Systematic Botany. 2003;**16**:1-18

[9] Orchard AE, Maslin BR. (1584) Proposal to conserve the name *Acacia* (Leguminosae: Mimosoideae) with a conserved type. Taxon. 2003;**52**:362-363

[10] Brummitt RK. Report of the Committee for Spermatophyta: 55. Proposal 1584 on *Acacia*. Taxon. 2004;**53**:826-829

[11] Luckow M, Hughes C, Schrire B, Winter P, Fagg C, Fortunato R, Hurter J, Rico L, Breteler FJ, Bruneau A, Caccavari M, Craven L, Crisp M, Delgado AS, Demissew S, Doyle JJ, Grether R, Harris S, Herendeen PS, Hernández HM, Hirsch AM, Jobson R, Klitgaard BB, Labat J, Lock M, MacKinder B, Pfeil B, Simpson BB, Smith GF, Sousa MS, Timberlake J, van der Maesen JG, Van Wyk AE, Vorster P, Willis CK, Wieringa JJ, Wojciechowski MF. *Acacia*: The case against moving the type to Australia. Taxon. 2005;**54**:513-519

[12] Kyalangalilwa B, Boatwright JS, Daru BH, Maurin O, Van der Bank M. Phylogenetic position and revised classification of *Acacia s.l.* (Fabaceae: Mimosoideae) in Africa,

including new combinations in *Vachellia* and *Senegalia*. Botanical Journal of the Linnean Society. 2013;**172**:500-523

[13] Ross JH. A conspectus of the African *Acacia* species. Memoirs of the Botanical Survey of South Africa. 1979;**44**:1-155

[14] Ward D. Population differentiation in a purported ring species, *Acacia karroo* (Mimosoideae). Biological Journal of the Linnean Society. 2011;**104**:748-755

[15] Archibald S, Bond WJ. Growing tall vs growing wide: Tree architecture and allometry of *Acacia karroo* in forest, savanna, and arid environments. Oikos. 2003;**102**:3-14

[16] Boon R. Pooley's Trees of Eastern South Africa: A Complete Guide. 2nd ed. Durban: Flora and Fauna Publications Trust; 2010

[17] Swartz PP. The correct name for *Acacia montana*. Bothalia. 2003;**33**:164-165

[18] Barnes RD, Filer DL, Milton SJ. *Acacia karroo* monograph and annotated bibliography. Tropical Forestry Papers. 1996;**32**:1-77

[19] Van Wyk B, Van Wyk P. Field Guide to Trees of Southern Africa. Cape Town: Struik Publishers; 1997

[20] Owen-Smith N, Cooper SM. Palatability of woody plants to browsing ruminants in a South African savanna. Ecology. 1987;**68**:319-331

[21] Teague WR. The rate of consumption of bush and grass by goats in a representative *Acacia karroo* Savanna community in the Eastern Cape. Journal of the Grassland Society of Southern Africa. 1989;**6**:8-13

[22] Miller MF. *Acacia* seed survival, seed germination and seedling growth following pod consumption by large herbivores and seed chewing by rodents. African Journal of Ecology. 1995;**33**:194-210

[23] Parker DM, Bernard RTF, Colvin SA. The diet of a small group of extralimital giraffe. African Journal of Ecology. 2003;**41**:245-253

[24] Van Vegten JA. Thornbush invasion in a savanna ecosystem in eastern Botswana. Vegetatio. 1983;**56**:3-7

[25] Trollope WSW. Controlling bush encroachment with fire in the savanna areas of South Africa. Proceedings of the Grassland Society of Southern Africa. 1980;**15**:173-177

[26] Du Toit PF. Bush encroachment with specific reference to *Acacia karroo* encroachment. Proceedings of the Grassland Society of Southern Africa. 1967;**2**:119-126

[27] Acocks JPH. Veld types of South Africa. Memoirs of the Botanical Survey of South Africa No 57. Pretoria: Botanical Research Institute; 1988

[28] Du Toit PF. A preliminary report on the effect of *Acacia karroo* competition on the composition and yield of Sweet Grassveld. Proceedings of the Grassland Society of Southen Africa. 1968;**3**:147-149

[29] Du Toit PF. The goat in a bush-grass community. Proceedings of the Grassland Society of Southern Africa. 1972;**7**:44-50

[30] Moore A, Van Niekerk JP, Knight IW, Wessels H. The effect of Tebuthiuron on the vegetation of the thorn bushveld of the Northern Cape: A preliminary report. Journal of the Grassland Society of Southern Africa. 1985;**2**:7-10

[31] Mucunguzi P. Effects of bruchid beetles on germination and establishment of *Acacia* species. African Journal of Ecology. 1995;**33**:64-70

[32] Brown NAC, de V. Booysen P. Seed coat impermeability in several *Acacia* species. Agroplantae. 1969;**1**:51-59

[33] O'Connor TG. *Acacia karroo* invasion of grassland: Environmental and biotic effects influencing seedling emergence and establishment. Oecologia. 1995;**103**:214-223

[34] Du Toit PF. *Acacia karroo* intrusion: The effect of burning and sparing. Proceedings of the Grassland Society of Southern Africa. 1972;**7**:23-27

[35] Mbalo BA, Witkowski ETF. Tolerance to soil temperatures experienced during and after the passage of fire in seeds of *Acacia karroo, A. tortilis* and *Chromolaena odorata*: A laboratory study. South African Journal of Botany. 1997;**63**:421-425

[36] Du Toit JCO, Sekwadi KP. Tebuthiuron residues remain active in soil for at least eight years in a semi-arid grassland, South Africa. African Journal of Range & Forage Science. 2012;**29**:85-90

[37] Harmse CJ, Kellner K, Dreber N. Restoring productive rangelands: A comparative assessment of selective and non-selective chemical bush control in a semi-arid Kalahari savanna. Journal of Arid Environments. 2016;**135**:39-49

[38] Belsky AT, Amundson RG, Duxbury JM, Riha SJ, Ali AR, Mwonga SM. The effects of trees on their physical, chemical, and biological environments in a semi-arid savanna in Kenya. Journal of Applied Ecology. 1989;**26**:1005-1024

[39] Scholes RJ, Archer SR. Tree-grass interactions in savannas. Annual Review of Ecology and Systematics. 1997;**28**:517-544

[40] Barnes RD. The African *Acacias*: A thorny subject. Southern African Forestry Journal. 2001;**190**:9-18

[41] Alcamo IE. Fundamentals of Microbiology. 6th ed. Sudbury, Massachusetts: Jones and Bartlett Publishers; 2001

[42] Taiz L, Zeiger E. Plant Physiology. 5th ed. Sunderland, Massachusetts: Sinauer Associates Inc. Publishers; 2010

[43] Salisbury FB, Ross CW. Plant Physiology. 4th ed. Belmont, California: Wadsworth Publishing Company; 1992

[44] Smit GN, Van Romburgh KSK. Relations between tree height and the associated occurrence of *Panicum maximum* in Sourish mixed Bushveld. African Journal of Range and Forage Science. 1993;**10**(3):151-153

[45] Kennard DG, Walker BH. Relationships between tree canopy cover and *Panicum maximum* in the vicinity of Fort Victoria. Rhodesia Journal of Agricultural Research. 1973;**11**:145-153

[46] Smit GN, Swart JS. Influence of leguminous and non-leguminous woody plants on the herbaceous layer and soil under varying competition regimes in mixed Bushveld. African Journal of Range and Forage Science. 1994;**11**:27-33

[47] Stuart-Hill GC, Tainton NN, Barnard HJ. The influence of an *Acacia karroo* tree on grass production in its vicinity. Journal of the Grassland Society of Southern Africa. 1987;**4**:83-88

[48] Pote J, Shackleton C, Cocks M, Lubke R. Fuelwood harvesting and selection in valley thicket, South Africa. Journal of Arid Environments. 2006;**67**:270-287

[49] Robbins CT, Mole S, Hagerman AE, Hanley TA. Role of tannins in defending plants against ruminants: Reduction in dry matter digestion? Ecology. 1987;**68**:1606-1615

[50] Bennett RN, Wallsgrove RM. Secondary metabolites in plant defence mechanisms. New Phytologist. 1994;**127**:617-633

[51] Barbehenn RV, Constabel CP. Tannins in plant–herbivore interactions. Phytochemistry. 2011;**72**:1551-1565

[52] Cannas A. Tannins: Fascinating but Sometimes Dangerous Molecules [Internet]. 2001. Available from: http://poisonousplants.ansci.cornell.edu/toxicagents/tannin.html [Accessed: March 29, 2017]

[53] Bartwal A, Mall R, Lohani P. Role of secondary metabolites and brassinosteroids in plant defense against environmental stresses. Journal of Plant Growth Regulation. 2013;**32**:216-232

[54] Dingaan MNV. Interpretation of the *Acacia karroo* Class, Southern Africa [Thesis]. Bloemfontein: University of the Free State; 2008

[55] Scholes RJ. Savanna. In: Cowling RM, Richardson DM, Pierce SM, editors. Vegetation of Southern Africa. Cambridge: Cambridge University Press; 1997. p. 258-277

[56] Low AB, Rebelo AG. Vegetation of South Africa, Lesotho and Swaziland. Pretoria: Department of Environmental Affairs and Tourism; 1996

[57] Rutherford MC, Westfall RH. Biomes of southern Africa: An objective categorization. Memoirs of the Botanical Survey of South Africa No 64. Pretoria: National Botanical Institute; 1994

[58] Rutherford MC, Mucina L, Lötter MC, Bredenkamp GJ, Smit JHL, Scott-Shaw CR, Hoare DB, Goodman PS, Bezuidenhout H, Scott L, Ellis F, Powrie LW, Siebert F, Mostert TH, Henning BJ, Venter CE, Camp KGT, Siebert SJ, Matthews WS, Burrows JE, Dobson L, van Rooyen N, Schmidt E, Winter PJD, du Preez PJ, Ward RA, Williamson S, Hurter PJ. Savanna Biome. In: Mucina L, Rutherford MC, editors. The Vegetation of South Africa,

Lesotho and Swaziland. Strelitzia No 19. Pretoria: South African National Botanical Institute; 2006

[59] Winterbach R, Bredenkamp GJ, Deutschländer MS, Mucina L. Preliminary syntaxonomic scheme of vegetation classes for the central Bushveld of South Africa. In: White PS, Mucina L, Leps JS, Van der Maarel E, editors. Proceedings IAVS Symposium. Uppsala: Opulus Press; 2000. p. 123-127

[60] Siebert SJ. Vegetation on the ultramafic soils of the Sekhukhuneland Centre of Endemism [Thesis]. Pretoria: University of the Pretoria; 2001

[61] Siebert SJ, Van Wyk AE, Bredenkamp GJ. The physical environment and major vegetation types of Sekhukhuneland, South Africa. South African Journal of Botany. 2002;**68**:127-142

[62] Breebaart L, Deutschländer M. The vegetation types and management units of Goedverwacht farm in the mixed bushveld of the Northern Province, South Africa. Koedoe. 1997;**40**:19-35

[63] Schmidt AG, Theron GK, Van Hoven W. The phytosociology and structure of vegetation near villa Nora, north-western Transvaal, South Africa. South African Journal of Botany. 1993;**59**:500-510

[64] Winterbach R. A phytosociological synthesis of *Acacia tortilis* communities in the north-western Savanna of South Africa [thesis]. Pretoria: University of Pretoria; 1998

[65] Brown LR, Bredenkamp GJ, Van Rooyen N. Phytosociological synthesis of the vegetation of the Borakalo nature reserve, north-West Province. South African Journal of Botany. 1997;**63**:242-253

[66] Brown LR, Bredenkamp GJ. The phytosociology of the southern section of Borakalo nature reserve, South Africa. Koedoe. 1994;**37**:59-72

[67] Coetzee BJ. A phytosociological classification of the Rustenburg nature reserve. Bothalia. 1975;**11**:561-580

[68] Bezuidenhout H. The major vegetation communities of the Augrabies falls National Park, northern cape. 1. The southern section. Koedoe. 1996;**39**:7-24

[69] Mucina L, Hoare DB, Lötter MC, du Preez PJ, Rutherford MC, Scott-Shaw CR, Bredenkamp GJ, Powrie LW, Scott L, Camp KGT, Cilliers SS, Bezuidenhout H, Mostert TH, Siebert SJ, Winter PJD, Burrows JE, Dobson L, Ward RA, Stalmans M, Oliver EGH, Siebert F, Schmidt E, Kobisi K, Kose L. Grassland biome. In: Mucina L, Rutherford MC, editors. The Vegetation of South Africa, Lesotho and Swaziland. Strelitzia No 19. Pretoria: South African National Botanical Institute; 2006

[70] Scholes RJ, Biggs R. A biodiversity intactness index. Nature. 2005;**434**:45-49

[71] O'Connor TG, Bredenkamp GJ. Grassland. In: Cowling RM, Richardson DM, Pierce SM, editors. Vegetation of Southern Africa. Cambridge: Cambridge University Press; 1997. p. 215-257

[72] Tainton NM. Veld Management in South Africa. Pietermaritzburg: University of Natal Press; 1999

[73] Matthews WS, Bredenkamp GJ, Van Rooyen N. The vegetation of the dry dolomitic regions of the North-eastern Mountain Sourveld of the Transvaal escarpment, South Africa. Phytocoenologia. 1992;**20**:467-488

[74] Burgoyne PM. Phytosociology of the north-eastern Transvaal high mountain grasslands [Thesis]. Pretoria: University of Pretoria; 1995

[75] Werger MJA. A phytosociological study of the upper Orange River valley. Memoirs of the Botanical Survey of South Africa. 1980;**46**:1-98

[76] Du Preez PJ, Bredenkamp GJ. Vegetation classes of the southern and eastern Orange free state (Republic of South Africa) and the highlands of Lesotho: Description of classes. Navorsinge van die Nasionale Museum. 1991;**7**:485-495

[77] Du Preez PJ. A syntaxonomical and synecological study of the vegetation of the south-eastern Orange Free State and related areas with special reference to Korannaberg [dissertation]. Bloemfontein: University of the Orange Free State; 1991

[78] Müller DB. Plantekologie van die Willem Pretorius-Wildtuin [dissertation]. Bloemfontein: University of the Orange Free State; 1986

[79] Robbeson RAJ. Phytosociology of northwestern KwaZulu-Natal [Thesis]. Pretoria: University of Pretoria; 1998

[80] Bezuidenhout H, Bredenkamp GJ, Theron GK. The vegetation of the Bd and Ea land types in the grassland of the western Transvaal, South Africa. South African Journal of Botany. 1993;**59**:319-331

[81] Coetzee JP, Bredenkamp GJ, Van Rooyen N. The phytosociology of the grasslands of the Ba and Ib land types in the Pretoria-Witbank-Heidelberg area. South African Journal of Botany. 1995;**61**:123-133

[82] Eckhardt HC, Van Rooyen N, Bredenkamp GJ. Plant communities and species richness of the *Agrostis lachnantha-Eragrostis plana* wetlands of northern KwaZulu-Natal. South African Journal of Botany. 1996;**62**:306-315

[83] De Frey WH. Phytosociology of the Mpumalanga high altitude grasslands [Thesis]. Pretoria: University of Pretoria; 1999

[84] Smit CM, Bredenkamp GJ, Van Rooyen N, Van Wyk AE, Combrinck JM. Vegetation of the Witbank nature reserve and its importance for conservation of threatened rocky Highveld grassland. Koedoe. 1997;**40**:85-104

[85] Mucina L, Rutherford MC, Palmer AR, Milton SJ, Scott L, Lloyd JW, van der Merwe B, Hoare DB, Bezuidenhout H, Vlok JHJ, Euston-Brown DIW, Powrie LW, Dold AP. Nama-Karoo Biome. In: Mucina L, Rutherford MC, editors. The Vegetation of South Africa,

Lesotho and Swaziland. Vol. 19. Strelitzia No 19. Pretoria: South African National Botanical Institute; 2006

[86] Palmer AR, Hoffman MT. Nama-karoo. In: Cowling RM, Richardson DM, Pierce SM, editors. Vegetation of Southern Africa. Cambridge: Cambridge University Press; 1997. p. 167-188

Ecological Biogeography of West Usambara Mountains: A Study on the Influence of Abiotic Factors to Spatial Distribution of Plant and Animal Species

Joel Loitu Meliyo, Kenneth F.G. Masuki,
Balthazar M. Msanya, Didas N. Kimaro and
Loth S. Mulungu

Additional information is available at the end of the chapter

http://dx.doi.org/10.5772/intechopen.72068

Abstract

West Usambara Mountains, Tanzania are famous for rich biodiversity and endemic species of plants and animals. Although there have been extensive studies on plant and animals species, little attention has been given on abiotic factors influencing their spatial distribution. Given rampant degradation of vegetation and associated consequence on biodiversity, knowledge on abiotic factors influencing distribution of species along the landscape become pertinent for conservation. A study was carried out to explore abiotic factors impacting plant and animals species distribution. Soil, landform and land use/cover were studied using grids of 20 m × 20 m using FAO Guidelines for Soil Profile and Habitat Descriptions. Soils were described, sampled for laboratory analysis. Spatial distribution of plant species were determined in the grids, and along the transects, every time estimating the percent cover and describing the habitat. Distribution of animal species was studied using both small mammals and rodent burrows as proxies. Trapping was done using different traps sizes, checking daily for animal caught, counting and re-trapping. Rodent burrows were estimated in same grids by examining a width of 0.5 m from end to end of grid and total number of burrows recorded. Determination of species distribution was done using GLM regression. Results show that species are influenced by elevation, which was common to both plants and animals. Topsoil soil depth was positive to plant species whereas hillshade, surface stones, cultivation and atmospheric temperature were negatively influencing plant species. Rock outcrops, surface stones and cultivation were positively influencing small mammals distribution. It is concluded that factors influencing distribution of small mammals are elevation, surface stones, rock outcrop and cultivation. Factors influencing plant species are elevation soil depth whereas cultivation, hillshade, surface stone and rock out crops negatively impact distribution. For conservation, it is recommended that the best steps are to stop human activities leading to depletion of plant species and accelerating soil erosion and allow for self-regeneration. Control of soil erosion strongly recommended as way of plant species re-establishment.

Keywords: abiotic factors, spatial distribution, small mammals, rodent burrows, landforming processes

1. Introduction

Usambara Mountains are famous Mountains ranges located at the Northeastern part of Tanzania. These are Block Mountains forming a series of 12 separate mountains described as the Eastern Arc Mountains (EAM). The EAMS stretches from Kenya the Taita hills through Udzungwa Mountains in southern highlands of Tanzania [1]. Most of the EAMs 11 separates mountains are found in Tanzania namely: North Pare, South Pare, West Usambara, East Usambara, Nguu, Nguru, Uluguru, Malundwe, Ukaguru, Rubeho, and Udzungwa [2]. The name 'Eastern Arc Mountains' (EAM) was coined as a suitable way of defining unique forests areas that comprises many rarely found plants' and animals' species [3].

Literature shows existence of intensive and extensive studies since 1800s that targeted flora and fauna throughout the individual mountains of the Eastern Arc. There are discoveries made on biological diversity for small mammals [4–7], larger mammals such as primates [8], carnivores [9], invertebrates [10], reptiles and amphibians [11], and birds [12–14]. The literature also indicates rich diversity of plants like angiosperms [15, 16]. So, far research works indicated the existence of over 100 species of birds, mammals, amphibians and reptiles, and also over 500 plants, and vast numbers of butterflies and millipedes [17].

These long-time research efforts that were made on the biodiversity treasures of the EAMs have led to the global recognition of EAMs as a very important global biological rich heritage [18], therefore, the EAMs have been designated as the 25th world's biodiversity hotspots [19, 20] and one of the World Wild Fund's Global 200 priority ecoregions [2]. Furthermore, the EAMs are not only important as the global biodiversity hotspots, but in Tanzania, the Eastern Arc forests are the source of 90% of water flowing for the hydroelectric power of the country. The forests are also the source of water for major cities including Dar es Salaam, Morogoro and Tanga all of which with human population of over 10 million people. The EAMs watersheds in Tanzania are also flowing waters through some of the National Parks including Udzungwa, Mikumi, Mkomazi, Saadani and Selous Game Reserve. In general, the EAMs are crucial for both as an ecological haven and also as socio-economic treasure of the country.

The Usambara Mountains are formed by two separate land massifs, the 'east' and 'west Usambara' Mountains, that are separated by a 4 km wide Lwengera River Valley. These Mountains share the uniqueness of the EAM characterised by the myriads of endemic flora and fauna: a rich biodiversity, and perhaps the most studied in the EAMs [19, 21, 22]. The biologists and ecologists studied well and established factors that governed high proportion of endemism and biodiversity in the Usambara Mountains and the entire EAMs series, linking biodiversity with long periods of tropical forest cover attributed to reliable rainfall onsets and patterns [21, 23]. However, the studies on abiotic factors influencing biotic forest dwellers were limited in the EAMs. The abiotic factors including climate, geology, soils, and landform characteristics are strong determinants of plants and animal distribution [24, 25].

The organism presence is expressed in the theory of the 'ecological niche of a species', defined by Hutchinson [26] to be 'a sum of all the environmental factors acting on the organism, in a region of n-dimensional hyperspace' of which each factor acting on the organism is one dimension within which a species can theoretically maintain a viable population. The multiple dimensions [26] include factors such as physical, chemical and biological parameters which set-up a niche with a range of prevailing conditions (landform, soils, temperature, rainfall) and resources (food, water, breeding sites, safe refuge) within which a species can persist. Knowing factors making the ecological niche are important because conservationists get to know the effects environment has on species and vice versa the effects species has on the environment. Previous studies [18–20] mainly covered species in their niche with little coverage how the environmental (abiotic) factors impact survival of the species. Lack of knowledge consequences is that endemic species of plants and animal in West Usambara Mountains and EAMs maybe threatened as abiotic factors are degrading to irreversible levels due to human interventions.

The aim of this study is to contribute the understanding of the influences the abiotic factors have on the biodiversity of West Usambara Mountains. It is expected that this will shed light to various actors on what ought be done to conserve the devastated rich biodiversity in west Usambara Mountains due changing abiotic factors including human activities such forest deforestation.

2. Methodology

2.1. Description of the study area

The West Usambara Mountains are 31 km wide and narrowed to the east 19 km wide and approximately 90 km long. The study area is a rectangle comprising part of adjoining plain, escarpment and part of plateau. The area is selected because it has diverse and unique characteristics that may help to understand how abiotic factors are influencing spatial distribution of plant and animals' species from hot dry plains across the escarpment to the cool plateau (**Figure 1**).

The study lies between latitude 4°30′ and 4°45′S and longitude 38°00′ and 38°45′E. It is located in a cold and warm dry zones of West Usambara Mountains. The plateau section receives annual precipitation of 1200 mm and less than 1000 mm for cold and warm plateau, respectively. The study area extends from the plain across a steeper escarpment both located in a

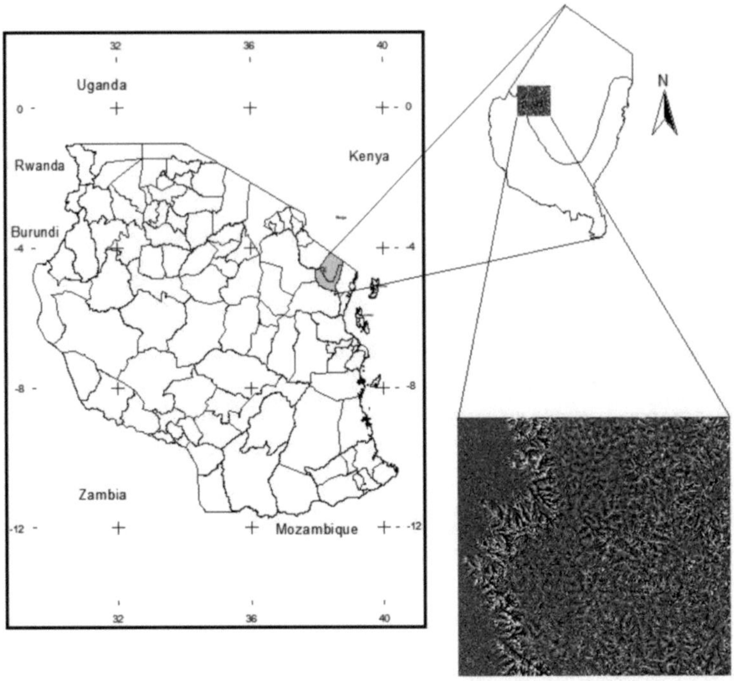

Figure 1. Location of biogeography study area, in West Usambara Mountains, Tanzania.

rain shadow side of the Usambara Mountains (plateau). The plain and escarpment receives annual precipitations of 400 and 800 mm, respectively. The average annual temperatures for plateau cold and warm zones are 14°C and 27°C, respectively, whereas the relative humidity is 70%. Temperatures in the plain range from 25 to 34°C per annum. The plateau grows diverse tropical crops and temperate fruits, while the plain has limited farming activities except for sisal (*Agave sisalana*).

2.2. Geology of West Usambara Mountains

West Usambara Mountains were formed by block faulting and repeated uplifting of Precambrian basement rocks between 180 and 290 million years ago [27, 28]. The geologic composition and variability in West Usambara Mountains depicts differences in terms of geology with regards to geomorphic position. The **plain** is mainly composed of duricrust calcareous yellow grey sand, whereas the geological rocks in the **escarpment** are gneisses mixed with undifferentiated granulites and distinctive bands of hornblende and pyroxenes. The geology in the **plateau** is mainly gneisses with leucocratic quartzo-feldspathic granulites and khondalites. There are few areas composed of recent alluvial materials like depressions and valley bottoms which are composed of mixed alluvial-fluvial materials [29].

2.3. Determination of landform and soil characteristics

Visual observation of satellite images, stereoscopic aerial photographs and orthophoto maps (numbers 9480410, 9480400, 9470410, and 9470400) was done. Also, visual analysis of topographic and geologic maps was done. The interpretation base-map was digitised to produce the georeferenced base-map used to guide field. Transect were made for augering for mapping soils and landforms by the methods by Dent and Young [30]. At each observation site, data on landform and soil morphological characteristics were examined and recorded. Landform units similar in parent material, relief, topography and soil morphological characteristics were considered to be similar and were accorded as mapping unit. Vegetation habitats and associated characteristics were determined and mapped. Identification of vegetation was according to the FAO Guidelines [31] in grids of 20 m × 20 m. Each observation site was geo-referenced by Global Positioning System (GPS). Representative soil profiles were dug in major soils, where description was done, and then soil samples were collected from natural horizons for laboratory analysis.

Also, landform analysis was done using ASTER Digital Elevation Model (DEM) which was carried out using ArcGIS 9.3 to derive continuous surfaces for elevation (m a.s.l.) (slope gradient (degrees), slope aspect (radians), slope length (m), and slope types (straight, convexity and concavity) and different types of land surface curvatures.

2.4. Exploring species distribution with abiotic characteristics

Habitat or vegetation description was done in grids of 20 m × 20 m by estimate vegetation cover percentage. Spatial distribution of animal was done using two approaches. One was by trapping small mammals at sites where augering and/or soil (profile sites) were made using grids of 20 m × 20 m. The data were collected twice between December 2009 and March 2013. Traps of different types and sizes were employed to capture diverse mammal species such Sherman live traps, local made wire cages (for bigger sized small mammals like squirrel, genetta) and the pitfall traps, which are 10-l plastic buckets. The total numbers of traps used were 300 of which 270 were Sherman, 15 wire cages and 15 pitfalls. The traps were arranged in lines each with 10 trapping stations placed 10 m apart and left open during the day and night for two consecutive nights [32]. Traps were inspected every morning to remove trapped animals and replace the bait. Peanut butter mixed with maize bran, roasted maize grains and sardines were used as bait. The trapped small mammals were counted and recorded.

The second approach was use of a proxy, which was the rodent burrow. Total rodent burrows were estimated on the landscape. Rodents' burrows were scanned within grids of 20 m × 20 m at width of 0.5 m from one end of grid, return until finished. A number of burrows was recorded and the nearest to the centre was opened to see if the animal were in or how recent the use has been. Atmospheric temperature were measured using the infrared thermometer topsoil temperature were done using thermal couple thermometer. The topsoil relative humidity was estimated using iButtons buried 20 cm for 24–26 hours recording both relative humidity (%) and temperature (degree Celsius).

2.5. Laboratory determination of soil properties and soil classification

Selected soil physical properties were determined in the field such as soil depth (cm). Chemical and soil texture were determined in the laboratory using methods by Page and Keeney [33] and Klute [34] respectively. Micronutrients (iron, manganese, copper zinc) were determined using Diethylenetriaminepenta-acetic acid (DTPA) according to Moberg [35]. The field and laboratory data were used to classified soils to level-2 of the FAO World Reference Base [36]. Although chemical soil properties were not used in modelling, it was used for soil classification. For modelling only topsoil depth and texture were used as input data.

2.6. Statistical analysis

The data was organised for multiple regression analysis. There were two dependent variables (plant cover (%) and total rodent burrows. The independent variable examined were 25, which were landform types, slope gradient (degrees), slope length (m), slope form (concave, convex, straight, compound), elevation (m a.s.l.), drainage, erosion type, rock outcrops and surface stones (number), slope aspect, hillshade (radians), slope curvature types (radians), soil depth (cm), soil texture (textural class), atmospheric temperature (degrees Celsius), topsoil (10 and 30 cm) temperature and topsoil (10 and 30 cm depth) relative humidity (%) were model input data. There was a total of 487 data entries collected. Categorical data such as textural class were given dummy number.

Abiotic factors explaining spatial distribution of plants and animals species were established by inputting 25 factors in a Generalised Linear Model, distribution family 'Gaussian' which is a multinomial for multiple dependent variables [37] applying a formula:

$$Y_i = \beta_0 + \beta_1 X_{1i} + \beta_2 X_{2i} + \beta_3 X_{3i} + \varepsilon_i \quad (1)$$

Where: Y_i = respondent (dependent) variables (plant cover, trapped animals/rodent burrows as a proxy); β_o = Intercept; $\beta_1 X_{1i} + \ldots \beta_3 X_{3i}$ = predictors or independent variables; ε_i = error term.

Using R software the GUI rattle [38]. Model validation was addressed by portioning the data. The 70% of the data was allocated for training while 30% was used to develop the model. Different runs were made first using all predictors then reduced or added examining the model goodness of fit by looking the null and residual deviance and Akaike information criteria (AIC), whereby a model with a smallest AIC and a narrower gap between null and residual deviance was opted as model explaining the factors influencing species distribution along the landscape. Multicollinearity, was tackled by keying or deleting weakly correlated variables serially in the model.

3. Results and discussion

3.1. Variability of landforms

The geologic characteristics and folding and faulting of the area have had significant impact on drainage line and river systems all along the plateau and escarpment. The faulting had

http://dx.doi.org/10.5772/intechopen.72068

stronger influence on the scarp formation where there are fault lines that led to vertical scarps, and/or hanging rocks. The dense drainage and rivers network which is identical to dissection of the plateau which mainly has been influenced by hydrological water flows suggesting that the dominant land forming process in the area has been denudation by moving water (erosion) in different forms such as mass wasting (landslides, mass movement), gully, rill and sheet. The different landform components and slope forms links well with landslide and in particular waste movement. The recent and past geomorphic processes influence not only the vegetation establishment but also the habitats and the diverse animals that occupy them. This assumption is similar to the description by Cottle [24] who pointed the relationship between the geology and biodiversity of both animal and plant species.

Figure 2 describes three different geomorphic units: the plain, escarpment and plateau. The units are congruent with the geology, and plateau is the largest and strongly dissected forming a complex landscape dominated by a network of ridges at different altitude levels or terraced ridging. The plateau is characterised into three distinctive terraced plateau levels differentiated by altitude, viz.: Plateau terrace level I (PTI) a landscape situated at the altitude over 2067 m a.s.l. (i.e. characterised by irregular, conical narrow cliffs or rock outcrop narrow (<10 m) summits with limited vegetation mostly due to shallow soil (<30 cm or rockiness). Plateau terrace level II (PTII) is composed of isolated ridges with altitude range of 1862–2067 m a.s.l., (i.e. characterised by narrow ridge summits with scarps, cliff or rock outcrops and or shallow soil depth) and hence limited vegetation. Plateau terrace level III (PTIII), situated at altitude range of 1657–1862 m a.s.l. (i.e. forming a continuum of low ridges

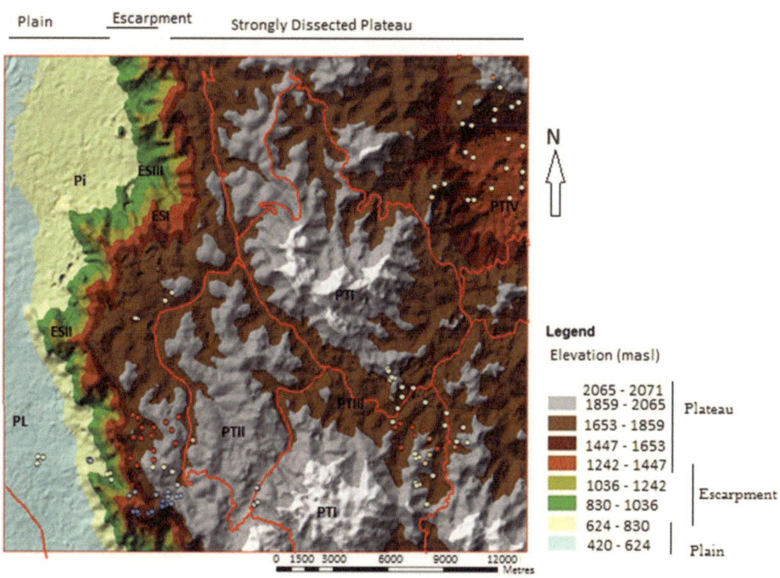

Figure 2. Landscape variation in the LEPUS project study area, West Usambara Mountains, Tanzania.

characterised by comparably broad crests with few localised rock outcrops at summits and/or upper slopes) and well covered with diverse dense vegetation where human did not clear for cultivation. There is a strong correction between plant species distribution, landform characteristics and soil characteristics particularly soil depth and quantity of gravel and/or stoniness or rockiness.

The soils in the plateau are diverse but are congruent with the landforms position on which they occur. The soils found on upper slopes and on their ridges' crests of PTI and PTII are dominantly **Regosols,** and **Lithic Leptosols.** The mid and lower slopes of PTI ridge crests are complexes of **Cutanic Acrisols** and **Cutanic Alisols**. The PTII plateau soils are **Ferralic Cambisols** on the upper slopes and **Cutanic Acrisols** and **Ferralic Cambisols** on the mid slopes [39]. The soils on ridge crests, upper and mid slopes of plateau PTIII are dominantly complexes of **Cutanic Alisols** and **Haplic Regosols**. The dominant soils in the lower slopes of PTIII are **Luvic Ferralic Phaeozems** while the dominant soils of the very narrow valley bottoms of plateau are **Mollic Fluvisols, Gleyic Fluvisols** and **Antrosols** [40].

The entire plateau is composed of aggregated micro and macro watershed with high potential for soil loss through erosion. The erosion hazard is attributed to the steep slopes; weak soil structure and poor agronomic practices whereby farmers cultivate at very steep slopes of over 45° without conservation measures. The soils of the area had overall poor fertility. One of the macronutrient phosphorus is very low below 4 mgP/kg soil which may affect uptake of others. Also, Ca, Mg and K are low in most soils. Micronutrients Fe and Mn are in very large quantities whereas Cu and Zn are within recommended critical levels. These soils are good for establishment of most vegetation and habitats. However, for food crops, which most small mammals are depending upon as food, the poor soil fertility which is leading to poor crops and in dry years no crops will soon bring in natural selection especially to animal species whereby those which will not be able to scramble for small amount of food will perish and those which will adapt to smaller amount and new food will survival. From residents of the area, there are already several species of gazelle and wild pigs, which are no longer, found in the Usambara because of poor habitats and possibly availability of food. Furthermore, it is important to note that due to the influence of elevation on temperature the plateau is colder than the low plains. There are even variations between valley bottoms, higher ridges and Mountains in the Plateau, and congruent to soil variation, there are vegetation distribution and hence forest dwellers. The explanation agree well with reported by Cottle [24] and research work by Valencia et al. [25] and Baltzer et al. [41] that soil type have a strong influence on spatial distribution of plant species.

Escarpment geomorphic unit indicates three levels of uplift, indicating tectonic cycles and it's characterised by steep slopes, canyons, cliffs and rocks with slope gradients of over 72°. There are colluvial foothills, and slope complexes with varied slopes from 3 to 60°. In certain locations, steep slopes over 60° with deep, shallow and rock soils were observed. Escarpment rises from the plain at 600 m a.s.l., to over 2000 m a.s.l. (**Figure 2**). Lower escarpment is characterised by colluvial/alluvial foot slopes, scattered foot ridges and talus slopes. Dominant soils in escarpment are complexes of **Mollic Leptosols, Lithic Leptosols, Cutanic Luvisols** and

Haplic Cambisols while the associated vegetation species are shrubs and large trees where soils are deep. In canyons, *Ficus* spp., have been observed and dense shrubs occupied by different animal species including primate, wild pigs and diverse small mammals [6].

The plain is the lowest geomorphic unit in the study area (**Figure 2**) divided into the upper rolling, rolling and gently undulating plain, characterised by hot temperatures, low rainfall and deep soils developed from Neogene/Miocene deposits. The dominant soils are complexes of **Fluvic Cambisols** and **Mollic Fluvisols** on the lower plain and complexes of **Mollic Leptosols, Cutanic Luvisols** on the upper rolling plain and **Haplic Umbrisols** on colluvio-alluvial fans [39]. Dominant abiotic factors prevailing in the plain are low rainfall and higher temperatures, which are supporting the sparse vegetation mainly woody shrubs and thickets. The diversity of animals is higher because the plain is an animal corridor from nearby Mkomazi National Park. There is also an extensive influence of humans including over grazing. Generally, climatic and soil factors are major determinant of spatial distribution of animal and plant species, which is similar to reports by Valencia et al. [25] and Baltzer et al. [41] the influence of soil types and landform characteristics on the distribution of trees.

3.2. Factor influencing diversity and species distribution in West Usambara Mountains

3.2.1. *Effects of slope gradient*

Slope gradient a measures of steepness of the landform surface [42], is presented in **Figure 3** in degrees, and it varied with geomorphic units. **Figure 3** shows variability from the plain, escarpment to the dissected plateau. It varies from <1° in the plain to over 64° at the escarpment.

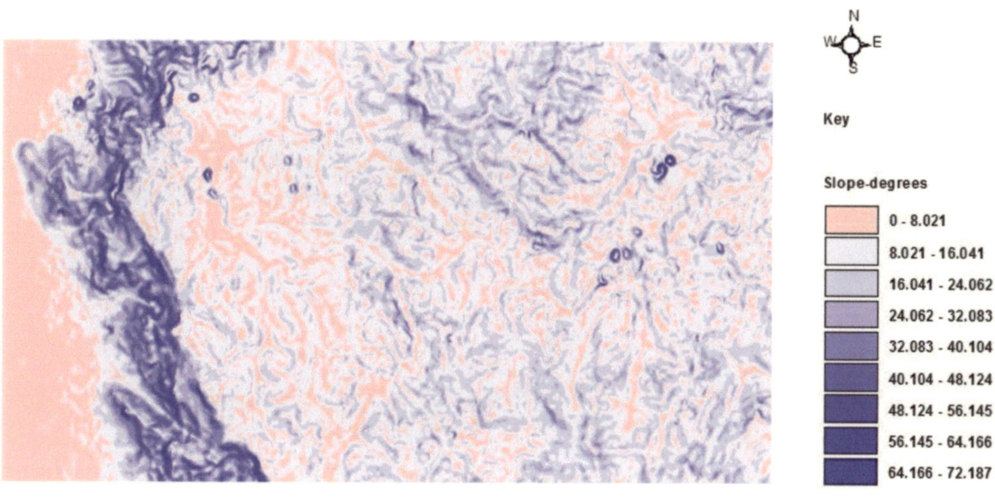

Figure 3. Slope gradient (degrees) variation along the landscape of West Usambara Mountains, Tanzania.

Very steep slopes are found in the escarpment and young landscape in the plateau. Gentle slopes are in valley bottoms and in the low plain. Due to the steep slope gradient, it is convincing that slope gradients and their types (**Figure 4**) are the strongest topographic attributes contributing significantly to landform forming processes by influencing speed, pathways of surface and subsurface water movement in the area. This conforms to report by Moore et al. [43] and Hutchinson [44] who indicated slope gradient to be among major factors of ecohydrology that influence overland flows. Similar results [45–47] indicate that slope influences water movement and landform forming processes including landslide in different parts of the globe. For instance, Zhou et al. [48] reported that most of the landslides that occurred in Hong Kong in 1993 took place on slope angles between 25 and 30°. Similarly, Mulders and Alexander [49] reported that areas with slope gradient of 35° and above have high likelihood of shallow landslides, whereas Fernandesa et al. [50] reported that in Brazil landslides are very common in slope angles between 37.1 and 55°.

Therefore, the west Usambara Mountain slopes of over 35°, mainly in escarpment and in plateau terraces PTI and PTII suggest that landslides could been one of the landforming processes in the past and still active given observable landslide scars in the area. The major slope forms are convex with excessive eroding and washing power hence exposing rocks to the surface, and/or concave with deposition manifested with deeper topsoils mainly none rocky or stony.

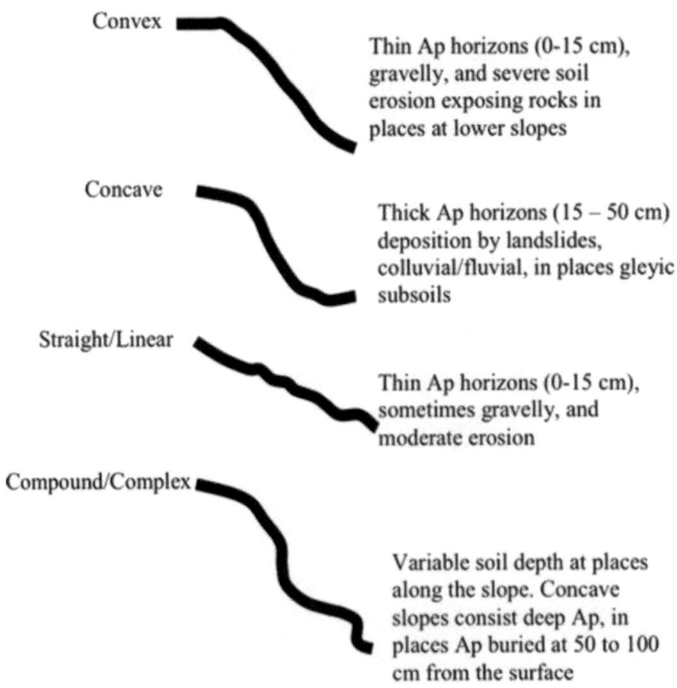

Figure 4. Dominant slope types in West Usambara Mountains, Tanzania.

3.2.2. Slope forms

Apart from slope gradient, slope forms (**Figure 4**) also vary in the area with convex slope form characterised with severe soil erosion (landslides, mass movement), hence has thin topsoils and in place stony and/or gravely surfaces and poor vegetation cover. Concave slope form are characterised as recipient sites with deep topsoils of over 50 cm in some places. Straight slope form although not apparent in the area, resembles convex in soil forming processes as they highly erodible, mainly washing by water. The complex slopes exhibit diverse properties depending on local area characteristics.

Slopes gradient and slope form have been found to strongly influence vegetation in Usambara Mountains. Dense vegetation is found in concave slopes with relatively deep soils and where moisture collects for longer times. This is in contrast to convex slopes with gravelly and shallows soils where there sparse vegetation and in many incases they harbour woody shrubs. It has been observed from this study that there are associations between vegetation establishment, and animals' species, which comply with studies by Njaka et al. [6] and Meliyo et al. [51]. This is in agreement with a work by Valencia et al. [25] who reported the influence of slope gradient and forms to trees distribution in South America.

3.2.3. Geomorphic processes

Active and dominant geomorphic processes operating and hence influencing spatial distribution of plant and animals along the landscapes is mainly water movement. Running water has effected by speed, dissolution and rock weathering which is facilitated by dissolved oxygen. Slope allows ponding or slow movement of water, which influences multiple physical chemical and biological processes.

Slope gradient dictates even the type of soil coverage in the area in terms of soil depth, textural compositions and even soil fertility status, and therefore establishment of vegetation and habitats and animal species. **Plate 1** shows geomorphogenetic processes mainly combined water and slope gradient.

Major processes are gravity mass movement, creep, rock fall, which are all accelerated by steep slopes and over saturation rainfalls. Geomorphomogenetic processes are active in the entire area, steep slope areas being highly prone to soil and debris removal whereas the low-lying areas are active sinks of seasonal varied textural materials. **Plate 1(a)–(d)** below indicates dominant morphogenetic processes shaping the landscape in the area. **Plate 1(b)** is mainly depositions of eroded materials of varied texture due to heavy rainfall.

3.3. Spatial distribution of animal and plant species along the landscape of West Usambara Mountains

Table 1 shows that plant and animal species occur across the three major geomorphic units. The diversity of species per landscape varies as has been reported by Meliyo et al. [51] and Njaka et al. [6]. The authors indicated that species specifically vegetation/habitat diversity

Plate 1. Morphogenetic processes shaping the landscape: (a) Mass movement where soils are eroded at wide portion of landscape down the slope after rainfall. (b) Shows deposition of varied rocks and boulders falling from upslope, (c) Shows old landslide scars prominent in the study area indicating the role played by landslide in the existing landform, and (d) Shows rock fall in the plateau, the practice very common along steep slopes over 25°, which is the case in the plateau and escarpment.

increases with increasing elevation. **Table 1** depicts that there were differences of abundance and types of vegetation cover/habitat and small mammals species found at different landscapes. Results show that there were fewer small mammals in the plain compared to the plateau. Although the number of small mammals increased with elevation (escarpment) the number in the plateau were more than the plain and the escarpment geomorphic units. Plant species or habitats in the low plain are mainly woody shrubs; thickets and even the scattered trees in some places are those salt tolerant species. This could be attributed to the fact that the plains are characterised by low sporadic rains and long period of droughts, that have led to development of sodic and saline soils, which only plants adapted to it could survive the hard shrubs and thickest.

The soils of the escarpment are mainly shallow, gravelly, stony and rocky and in some places just rocky land without soil. This implies that many plant species grow with difficulties. In few areas with deep soils, there were large trees and dense none thorny shrubs, well established compared to the drier plain. Moisture availability also could be a factor-segregating species distribution. **Table 1** further shows that some animal species are located only in one geomorphic unit and not the other. For instance, *Dwarf Mangoose*, *Genetta genetta* and Squirrels

http://dx.doi.org/10.5772/intechopen.72068

were just found in the plains, although different elevations. This could be attributed to the characteristics of the niche, which encompasses food availability, breeding places and or climate adaptation.

Landscape	Elev (m a.s.l.)	Slope (%)	Plant species/habitats	Trapped small mammals	Type small species
UP	480	7	WST	7	Acomys
UP	480	7	WST	1	*Dwarf Mongoose*
UP	480	7	WST	1	Squirrel
RP	615	15	WST	1	*Genetta genetta*
RP	615	12	WST	3	Squirrel
RP	615	15	WST	1	Acomys
RP	615	15	WST	1	Praomys
LE	830	30	WSST	3	Aethomys
LE	830	56	WSST	1	Acomys
UE	1350	80	WSST	7	Aethomys
UE	1350	55	WSST	1	Grammomys
UE	1350	55	WSST	1	Lophuromys
UE	1350	55	WSST	1	Mastomys
UE	1350	55	WSST	1	Otomys
UE	1350	55	WSST	1	Praomys
P	1850	23	CSST	1	Aethomys
P	1850	23	CSST	3	Crocidura
P	1850	90	CSST	24	Grammomys
P	1880	90	CSST	28	Lophuromys
P	1740	90	CSST	41	Mastomys
P	1740	70	CSST	4	Mouse legeda
P	1860	90	CSST	52	Praomys
P	1860	90	CSST	2	Rattus
P	1860	70	CSST	2	Shrew

UP, undulating plain; RP, rolling Plain; LE, lower escarpment; UE, upper escarpment; P, plateau; WST, woody shrubs and thickets; WSST, woody shrubs+surface stones; CSST, cropland+shrubs+surface stones.

Table 1. Spatial distribution of plant and small mammal species in West Usambara Mountains.

Plant species in the plateau are mainly plantation forest, where *Pine* spp., *Eucalyptus* spp., and *Camphor* spp., are planted for timber, project. Before 1980 most of the mountains and hills of the study area were treeless, due to deforestation followed after independence 1961, which then cleared thousands of forests for obtain farmland. Deforestation was followed by severe soil erosion, which the Tanzania Government intervened by formulating a project Soil Erosion Control and Agroforestry Project (SECAP) [52], which promoted tree plants and agroforestry. Hence individuals established tree woodlots in places where was stricken by soil erosion and landslides. However, in the plateau there are still few, small pockets of natural forest such Magamba Nature Reserve. That kind of forests remains, are the pockets harbouring the natural rich biodiversity of plant and animal species, west Usambara Mountains.

3.4. Abiotic factors explaining spatial distribution of species along the landscape

Table 2 and **Figure 5** present results that indicates that factors influencing spatial distribution of small mammals were elevation ($p < 0.001$), surface stones ($p < 0.001$), rock outcrop and cultivation (%) ($p < 0.05$) and slightly surface curvature (profile and cross) which is negatively influencing species spread across the landscape ($p < 0.1$). The atmospheric temperature and

Coefficients	Estimate	Std. error	t-value	Pr(>|t|)
(Intercept)	−5.8547295	1.2733596	−4.598	5.49e-06***
Elevation (m a.s.l.)	0.0024506	0.0006312	3.883	0.000118***
Rock_Out_crop	0.4354978	0.1870275	2.329	0.020307*
Surface_Stone	1.3789108	0.3826639	3.603	0.000347***
Slope_Aspect	0.0011544	0.0011295	1.022	0.307281
Profile_Curv	0.3181454	0.1881989	1.690	0.091599.
Cross_Curv	−0.3193107	0.1882792	−1.696	0.090558.
Soil_Depth1	−0.0175444	0.0158829	−1.105	0.269894
Soil_Depth3	−0.0055701	0.0150322	−0.371	0.711143
Cultivated_Area	0.0268368	0.0136358	1.968	0.049641*
Atmospheric T°C	0.0191391	0.0170757	1.121	0.262929

Significant codes: 0 '***' 0.001 '**' 0.01 '*' 0.05 '.' 0.1 ' ' 1.
Null deviance: 3956.1 on 481 degrees of freedom.
Residual deviance: 3026.9 on 471 degrees of freedom.
One observation deleted due to missingness.
AIC: 2277.5.

Table 2. Factors influencing distribution of small mammal species along the landscape of West Usambara Mountains, Tanzania.

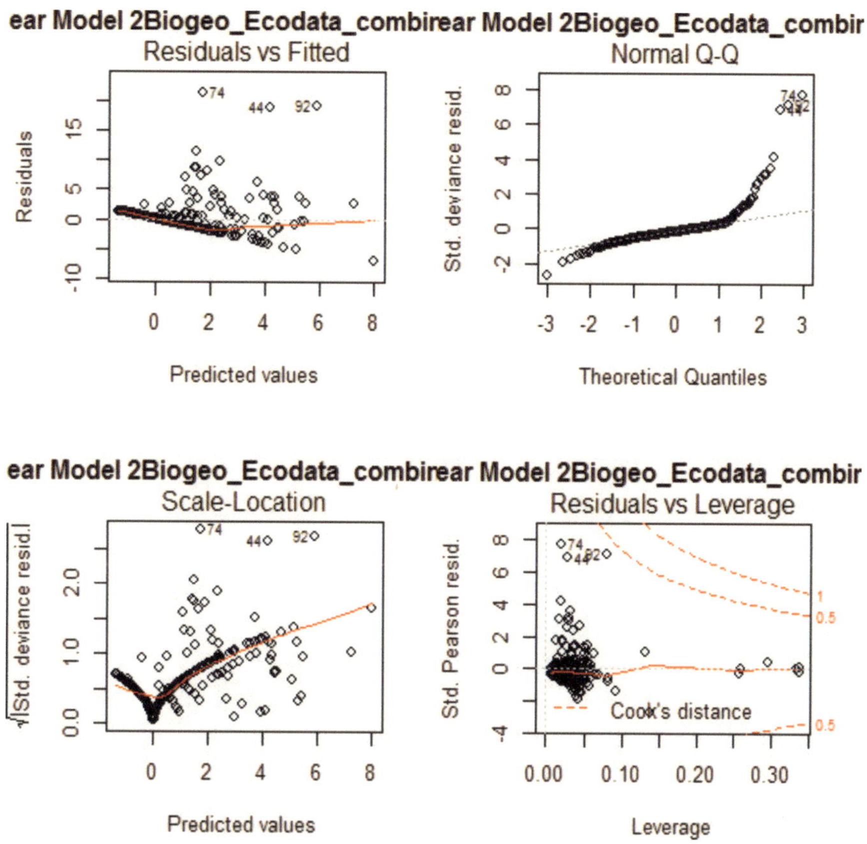

Figure 5. Plot of deviance residuals from GLM fitted to establish factors influencing small mammal species distribution along the landscape of West Usambara Mountain, Tanzania.

many other factors that were considered did not work out to be predictors. These results are congruent with field observation presented in **Table 1**, which show that the number of small mammals trapped was high in the plateau (higher elevation) than in the escarpment and the plain.

The results are too supported by those reported by Njaka et al. [6] who indicated that small mammal abundance increased with increasing elevation. The increase of small mammals with elevation could be attributed to favourable climatic conditions such occurrence of rainfall and reduced temperature compared to the low plain characterised by hot temperature and drought. Cultivation of food crops in the higher plateau due to comparably higher rainfall and cool temperatures serves as the habitat and continuous food supply which in short while attracts species but in future lead to extinction of those which do desire human interactions. The above account is in agreement with Ward et al. [53] who indicated that distribution of species is influenced by a large number of abiotic factors like environmental stability, habitat heterogeneity and relevance [54] and ecosystem production. It is also true that surface stones and rock outcrops offer small mammal both safe havens and breeding places. This implies that surface stones and rock outcrops create microhabitats that are hardly accessible to disturbances due to other organisms mostly humans. All predictors had positive coefficients, which show positive correlation of the independent variables to dependent ones. Surface stones and rock outcrops are microhabitats with unique characteristics, which influence spatial distribution of the small mammals particularly in the plateau [51]. The results also show that surfaces stones make microhabitat which influence abundance and hence distribution small mammals along the landscape, and the high population of small species at the plateau may be attributed to microhabitat as well as stable food supply associated to food crop cultivation. The results also agreed well with field data which show that few species were captured per trap station in the plain than the plateau, and also the plant species richness and diversity are greater in the plateau than the plain. Similar results were reported by Meliyo et al. [51] and Njaka et al. [6]. The results are also in agreement with those presented by Hastie et al. [55] who indicated that microhabitat influences local density of species and their spatial distribution, however, our results could not show that temperature and moisture to be important drivers of spatial distribution in the study area.

Table 3 and **Figure 6** present results depicting abiotic factors influencing spatial distribution of plant species along the landscape of West Usambara Mountains. The results show that elevation (metres above sea level), top soil depth and cultivation practices are major determinants of spatial distribution of plant species and they are statistically significant ($p < 0.001$) predictors, although cultivation practices are negatively influencing plant distribution i.e. vegetation clearing.

Other negatively influencing factors of statistical significance are hillshade ($p < 0.01$) and surface stones ($p < 0.05$) and atmospheric temperature in degrees Celsius (p < 0.05) (**Table 3** and **Figure 6**). Our results are in agreement with findings by Chen et al. [56] who studied factors affecting the distribution of pant species in Hainan Island, China, and reported many factors including elevation, soils, rainfall and human disturbances.

Similarly, our results on spatial distribution of plant species in the west Usambara Mountains are supported finding by Trigas et al. [57] who reported an increase in proportion of plant species endemism with increasing elevation of Cretan Mountain that could only be explained by elevation-driven ecological factors. Ecologically, there are many factors coming into play, including temperature, rainfall, soils, and where human disturbances occur particularly deforestation, these leading to plant species extinctions in some area in the world [56].

3.5. Human influence on distribution of small mammals and plant species in West Usambara Mountains

Table 3 and **Figure 7** present results of the influence of human being on small mammals and plant species. History shows that once the Usambara was covered by natural forests where diverse plants and myriads of small mammals were living in none-disturbed habitats. In 1980, most of the land has been cleared, and **Table 3** shows a strong statistical significance influence of cultivation on plant species. Other studies [58], compared biological diversity,

Coefficients	Estimate	Std. error	t-value	Pr(>|t|)
(Intercept)	−6.905e+01	1.838e+01	−3.757	0.000194***
Elevation (m a.s.l.)	6.761e-02	6.124e-03	11.041	<2e-16***
Slope_length	1.199e-01	1.092e-01	1.098	0.272673
Rock_Out_crop	−3.288e+00	1.694e+00	−1.941	0.052846.
Surface_Stone	−8.595e+00	3.445e+00	−2.495	0.012957*
Slope_Aspect	7.467e-03	1.105e-02	0.676	0.499410
Hillshade	−1.167e-01	3.542e-02	−3.295	0.001061**
Profile_Curv	3.692e+01	2.340e+01	1.577	0.115370
Plan_Curv	8.933e-04	1.235e-03	0.723	0.469747
Cross_Curv	2.419e+01	1.210e+01	1.999	0.046196*
General_Curv	−1.420e+01	1.037e+01	−1.369	0.171794
Longit_Curv	−3.325e+01	1.754e+01	−1.896	0.058643.
Tanget_Curv	−1.355e+01	2.430e+01	−0.557	0.577536
Soil_Depth1	7.674e-01	1.437e-01	5.341	1.46e-07***
Soil_Depth2	−5.494e-02	1.149e-01	−0.478	0.632629
Soil_Depth3	−1.241e-01	1.356e-01	−0.915	0.360555
Cultivated_area	−6.209e-01	1.239e-01	−5.014	7.64e-07***
Atmospheric T°C	−3.463e-01	1.634e-01	−2.119	0.034588*
Soil_Temper10cm	−2.299e-02	9.084e-02	−0.253	0.800349
Relat_Hum10cm	−2.412e-02	1.047e-01	−0.230	0.817888
Relat_Hum30cm	1.308e-01	1.346e-01	0.972	0.331514

Significant codes: 0 '***' 0.001 '**' 0.01 '*' 0.05 '.' 0.1 ' ' 1.
Null deviance: 360945 on 481 degrees of freedom.
Residual deviance: 233953 on 460 degrees of freedom.
AIC: 4395.

Table 3. Factors influencing distribution of plant species along the landscape of West Usambara Mountains, Tanzania.

indicated that human population growth has negatively affected natural resources, and hence biodiversity (**Figure 7**). Some hotspots such as eroded lands, deforested areas, dried water sources and undesirable tree species have been identified as creation of mankind, which are inversely related with rich biological diversity. Many animals (small and large) have been killed for food or because they destroy food crops planted. The consequences have been decline over time in per capita food production and increasing food insecurity and poverty, which are accelerating degradation of biodiversity of both plant and animal [58], particularly in the western facing drier West Usambara Mountains.

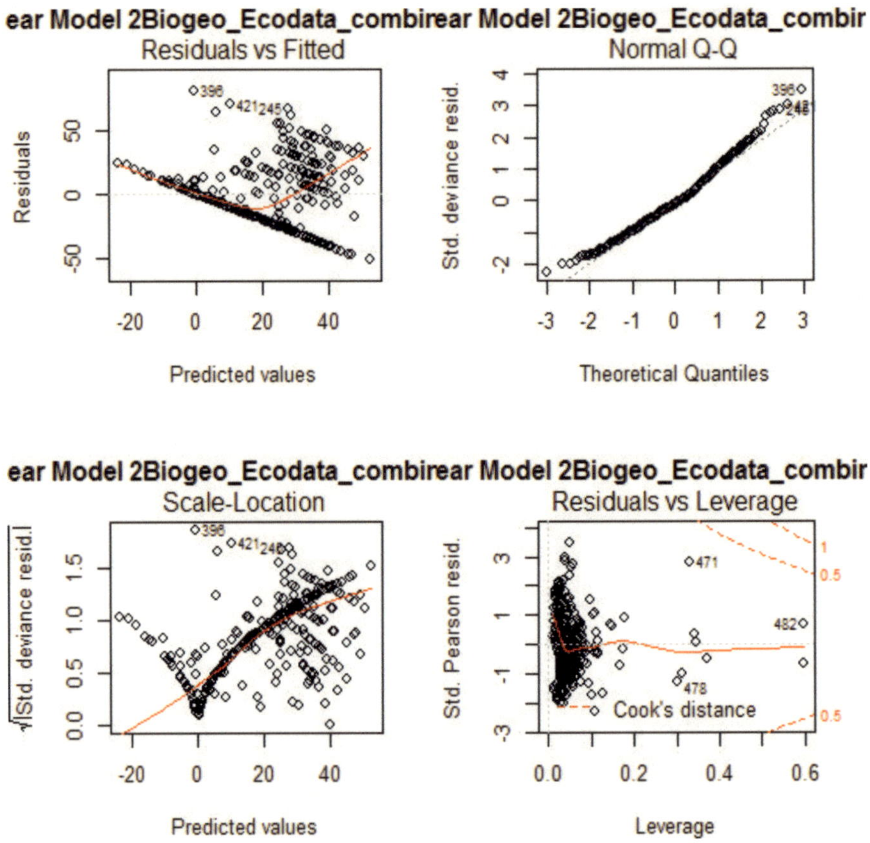

Figure 6. Plot of deviance residuals from GLM fitted to establish factors influencing plant species distribution along the landscape of West Usambara Mountain, Tanzania.

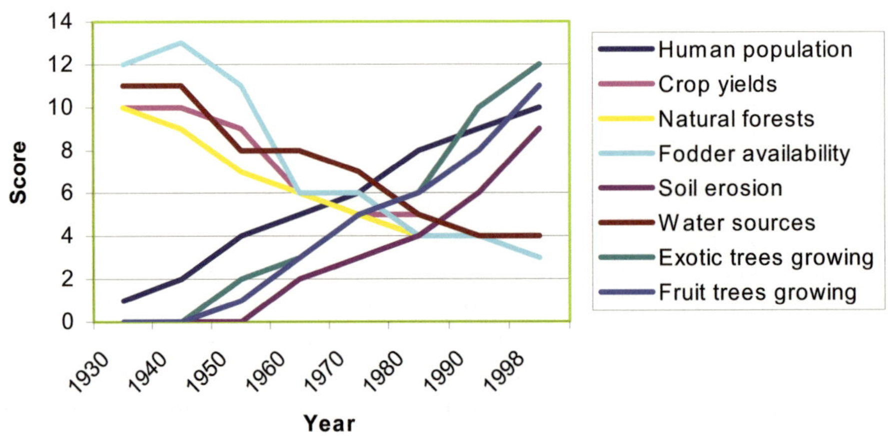

Figure 7. Historical account of human population pressure influence to the natural resources degradation West Usambara Mountains of Tanzania (Source: Meliyo et al. [58]).

4. Conclusions and recommendations

4.1. Conclusions

The study area was heterogeneous site composed of plain, escarpment and strongly dissected plateau in West Usambara Mountains, with different geology, landforms and soils types. The three geomorphic units have different plant and animal species – plain with shrubs and thickets which decline with increasing elevation while trees increase with increasing elevation from upper part of the escarpment to mountainous in the plateau. The geomorphological setting of the area has had the influence on plant and animal (small mammal) species spatial distribution. Abiotic factors explaining spatial distribution of plant species are landform characteristics including elevation, slope gradient and topsoil soil depth which have positive coefficients indicating that as factors increases so do plant species. There are factors negatively hindering spatial distribution including cultivation, which involve vegetation clearing, atmospheric temperatures and surface stones. Factors influencing spatial distribution of small mammals (animals) are elevation, surface stones, rock outcrop and cultivation. These factors signify favourable atmosphere, safe havens and food availability for small mammal to flourish. Most of the factors influencing both plant and animal species apart from cultivation, which involves deforestation or vegetation clearing are natural.

4.2. Recommendations

The factors influencing spatial distribution of plant and animal species in West Usambara Mountains have been established which are natural: elevation, atmospheric temperature, soil depth and slope gradients. The only factor, which is manmade, is cultivation which is negatively related to plant species but encourages spatial distribution of small mammals. Therefore, for conservation purposes, it is recommended that the best undertaking is stopping human activities leading to depletion of plant species and allow for self-regeneration. Facilitating control of erosion in steep slope areas is also recommended as way of plant species re-establishment. Land clearing for farming need to be control and establish a balance between human and ecologically acceptable land clearing considering that both farming and nature conservation are needed.

Acknowledgements

This project was funded by VLIR, Belgium, Ministry of Agriculture Food Security and Cooperative, Tanzania, and African Highland Initiatives (AHI). The authors would like to acknowledge many people in Lushoto District who facilitated in different ways the work including the Lushoto District Executive Director, District Agricultural and Livestock Development Officer and The Sebastian Kolowa Memorial University where the team camped for years.

Author details

Joel Loitu Meliyo[1*], Kenneth F.G. Masuki[2], Balthazar M. Msanya[3], Didas N. Kimaro[4] and Loth S. Mulungu[5]

*Address all correspondence to: joelmeliyo15@gmail.com

1 Selian Agricultural Research Institute, Arusha, Northern Tanzania

2 International Maize and Wheat Improvement Center (CIMMYT), Selian Agricultural Research Institute, Arusha, Tanzania

3 Department of Soil and Geological Sciences, Sokoine University of Agriculture, Morogoro, Tanzania

4 Department of Agricultural Engineering and Land Planning, Sokoine University of Agriculture, Morogoro, Tanzania

5 Pest Management Centre, Sokoine University of Agriculture, Morogoro, Tanzania

References

[1] Lovett JC, Wasser SK. Biogeography and Ecology of the Rain Forests of Eastern Africa. UK: Cambridge University Press; 1993

[2] Tabor K, Burgess ND, Mbilinyi BP, Kashaigili JJ, Steininger MK. Forest and woodland cover and change in Coastal Tanzania and Kenya, 1990 to 2000. Journal of East African Natural History. 2010;**99**(1):19-45

[3] Lovett JC. Eastern tropical African centre of endemism: A candidate for World Heritage Status? Journal of the East African Natural History Society. 1998;**87**:359-366

[4] Carleton MD, Stanley WT. Review of the *Hylomyscus denniae* complex (Rodentia: Muridae) in Tanzania, with description of a new species. Proceedings of the Biological Society of Washington. 2005;**118**:619-646

[5] Makundi RH, Massawe AW, Mulungu LS. Breeding seasonality and population dynamics of three rodent species in the Magamba Forest Reserve, Western Usambara Mountains, north-east Tanzania. African Journal of Ecology. 2006;**45**:17-21

[6] Njaka AR, Kimaro DN, Kihupi NI, Mulungu LS, Leirs H, Msanya BM, Deckers JA, Gulinck H. Vegetation habitats and small mammals in a plague endemic area in Western Usambara Mountains, Tanzania. Tanzania Journal of Health Research. July 2014;**16**(3): 1-12. DOI: 10.4314/thrb.v16i3.6

[7] Stanley WT, Goodman SM, Newmark WD. Small mammal inventories in the East and West Usambara Mountains, Tanzania. 1. Study areas, methodologies, and general results. In: Stanley WT, editor. Studies of Montane Vertebrates of Tanzania. Fieldiana Life and Earth Sciences. Chicago: Field Museum of Natural History. 2011;**4**:1-17

[8] Rovero F, Marshall AR, Jones T, Perkin A. The primates of the Udzungwa Mountains: Diversity, ecology and conservation. Journal of Anthropological Sciences. 2009;**87**:93-126

[9] De Luca DW, Mpunga NE. Small carnivores of the Udzungwa Mountains: Presence, distributions and threats. Small Carnivore Conservation. 2005;**32**:1-7

[10] Hoffman RL. Biogeography of east African montane forest millipedes. In: Lovett JC, Wasser SK, editors. Biogeography and Ecology of the Rain Forests of Eastern Africa. Cambridge, United Kingdom: Cambridge University Press; 1993. pp. 103-115

[11] Menegon M, Doggart N, Owen N. The Nguru Mountains of Tanzania, an outstanding hotspot of herpetofaunal diversity. Acta. Herpetologica. 2008;**3**:107-127

[12] John JRM, Kabigumil JDL. The use of bird species richness and abundance indices to assess the conservation value of exotic Eucalyptus plantations. Ostrich: Journal of African Ornithology. 2011;**82**(1):27-37

[13] Lens L, Dongen DV, Norris K, Githuri M, Matthysen E. Avian persistence in fragmented rainforest. Science. 2002;**298**:1236-1238

[14] Newmark WD. A 16-year study of forest disturbance and understory bird community structure and composition in Tanzania. Conservation Biology. 2006;**20**:122-134

[15] Cordeiro NJ, Howe HF. Forest fragmentation severs mutualism between seed dispersers and an endemic African tree. Proceedings of the National Academy of Sciences of the United States of America. 2003;**100**:14052-14056

[16] Lovett JC. Altitudinal variation in large tree community associations on the West Usambara Mountains. In: Hedberg I, Persson E, editors. Research for Conservation of Tanzania Catchment Forests. Uppsala: Uppsala Universitet; 1990. pp. 48-53

[17] EAMCEF (Eastern Arc Mountains Conservation Endowment Fund). Baseline Information for the Eight Nature Reserves and Udzungwa Mountains National Park in the Eastern Arc Mountains of Tanzania. Final Report produced by FORCONSULT. Morogoro, Tanzania: EAMCEF 2012

[18] BirdLife International. Biodiversity Status and Trends Report for the Eastern Arc Mountains and Coastal Forests of Kenya and Tanzania Region, 2012. Nairobi: BirdLife International – Africa Partnership Secretariat; 2013

[19] Burgess ND, Butynski TM, Cordeiro NJ, Doggart NH, Fjeldsa J, Howell KM, Kilahama FB, Loader SP, Lovett JC, Mbilinyi B, et al. The biological importance of the Eastern Arc Mountains of Tanzania and Kenya. Biological Conservation. 2007;**134**(2):209-231

[20] Mbilinyi BP, Malimbwi RE, Shemwetta DTK, Songorwa ZE, Katani JZ, Kashaigil J. Forest area baseline for the Eastern Arc Mountains. Conservation and Management of the Eastern Arc Mountain Forests Project, Forestry and Beekeeping Division, Ministry of Natural Resources and Tourism, United Republic of Tanzania, Dar es Salaam, Tanzania; 2006

[21] Conte CA. Highland Sanctuary: Environmental History in Tanzania's Usambara Mountains. Athens, Ohio: Ohio University Press; 2004

[22] Hubeau M. Land use and human activity patterns in relation to the plague disease in the West Usambara Mountains, Tanzania [MSc Dissertation]. Leuven, Belgium: Katholieke Universiteit; 2010. 97pp

[23] Lovett JC. Temperate and tropical floras in the mountains of eastern Tanzania. Opera Botanica. 1993;**121**:217-227

[24] Cottle R. Linking Geology and Biodiversity. Peterborough: English Nature Report No. 562; 2004

[25] Valencia R, Foster RB, Villa G, Condit R, Svenning J-C, Hernández C, Romoleroux K, Losos E, Magard E, Balslev H. Tree species distributions and local habitat variation in the Amazon: Large forest plot in eastern Ecuador. Journal of Ecology. 2004;**92**:214-229

[26] Hutchinson GE. Concluding remarks. Cold Spring Harbor Symposia on Quantitative Biology. 1957;**22**:415-427

[27] Griffiths CJ. The geological evolution of East Africa. In: Lovett JC, Wasser SK, editors. Biogeography and Ecology of the Rain Forests of Eastern Africa. UK: Cambridge University Press; 1993. pp. 9-21

[28] Hamilton AC. The place and the problem. In: Hamilton AC, Bensted-Smith R, editors. Conservation in the East Usambara Mountains, Tanzania Forest. Cambridge: IUCN; 1989. pp. 29-33

[29] Geological Survey Division. Geological Survey of Tanganyika, Lushoto. Quarter degree sheet No. 109. Dodoma, Tanzania: Geological Survey Division; 1963

[30] Dent D, Young A. Soil Survey and Land Evaluation. London: George Allen and Unwin; 1981

[31] Di Gregorio A. Land Cover Classification System: Classification Concepts and User Manual. Rome: Food and Agriculture Organization of the United Nations; 2005

[32] Mulungu LS, Makundi HR, Massawe AP, Machangu RS, Nsajigwa EM. Diversity and distribution of rodent and shrew species associated with variations in altitude on Mount Kilimanjaro, Tanzania. Mammalia. 2008;**72**:178-185

[33] Page AL, Miller RH, Keeney DR, editors. Methods of Soil Analysis, Part 2. 2nd ed., ASA, SSSA Monograph no. 9, Madison, Wisconsin: America Society of Agronomy, Inc., Soil Science Society of America, Inc.; 1982

[34] Klute A, editor. Methods of Soil Analysis. Physical and Mineralogical Methods. Madison, Wisconsin: America Society of Agronomy, Inc., Soil Science Society of America, Inc.; 1986

[35] Moberg JP. Soil Analysis Manual. Copenhagen, Denmark: KVL; 2000. 133pp

[36] WRB. World Reference Base for Soil Resources 2014; International Soil Classification System for Naming Soils and Creating Legends for Soil Maps. World Soil Resources Reports, 106. Rome, Italy: Food and Agriculture Organization of the United Nations; 2014

[37] McCulloch CE, Searle SR, Neuhaus JM. Generalized, Linear, and Mixed Models. 2nd ed. Hoboken, New Jersey: John Wiley & Sons, Inc.; 2008

[38] Williams GJ. Rattle: A Graphical User Interface for Data Mining in R. 2009b. URL: http://cran.r-project.org/package=rattle. R package version 2.4.55

[39] Massawe BHJ. Landform and soil analysis for predicting distribution of plague reservoirs and vectors in Mavumo Area, Lushoto District, Tanzania. Dissertation for Award of MSc Degree at Sokoine University of Agriculture, Morogoro, Tanzania; 2011

[40] Kimaro DN, Msanya BM, Meliyo JL, Hieronimo P, Mwango S, Kihupi NI, Hubert H, Deckers JA. Anthropogenic soils and land use patterns in relation to small mammal and flea abundance in plague endemic area of Western Usambara Mountains, Tanzania. Tanzania Journal of Health Research. July 2014;**16**(3):1-14

[41] Baltzer JL, Davies SJ, Noor NSMd, Kassim AR, LaFrankie JV. Geographical distributions in tropical trees: Can geographical range predict performance and habitat association in co-occurring tree species? Journal of Biogeography. 2007;**34**:1916-1926

[42] Zhang X, Drake NA, Wainwright J, Mulligan M. Comparison of slope estimates from low resolution DEMs: Scaling issues and a fractal method for their solution. Earth Surface Processes and Landforms. 1999;**24**:763-779

[43] Moore ID, Grayson RB, Landson AR. Digital terrain modelling: A review of hydrological, geomorphological, and biological applications. Hydrological Processes. 1991;**5**:3-30

[44] Hutchinson MF. Interpolating mean rainfall using thin plate smoothing spines. International Journal of Geographical Information Systems. 2007;**9**:385-404

[45] Gao J. Identification of topographic settings conducive to landsliding from DEM in Nelson County, VA, USA. Earth Surface Processes and Landforms. 1993;**18**:579-591

[46] Furian S, Barbiero L, Boulet R. Organization of the soil mantle in tropical southeastern Brazil (Serra do Mar) in relation to landslides processes. Catena. 1999;**38**:65-83

[47] Lopes JAU. The role of landslides in the landscape evolution: Theoretical and practical aspects. In: Pan-American Symposium on Landslides, Rio de Janeiro; 1997. pp. 91-100

[48] Zhou CH, Lee CF, Li J, Xu ZW. On the spatial relationship between landslides and causative factors on Lantau Island, Hong Kong. Geomorphology. 2002;**43**:197-207

[49] Mulder T, Alexander J. Abrupt change in slope causes variation in the deposit thickness of concentrated particle-driven density currents. Marine Geology. 2001;**175**:221-235

[50] Fernandesa NF, Guimara RF, Gomes RAT, Vieira BC, Montgomery DR, Greenberg H. Topographic controls of landslides in Rio de Janeiro: Field evidence and modelling. Catena. 2004;**55**:163-181

[51] Meliyo JL, Massawe BHJ, Msanya BM, Kimaro DN, Hieronimo P, Mulungu LS, Kihupi NI, Deckers JA, Gulinck H, Leirs H. Landform and surface attributes for prediction of rodent burrows in the Western Usambara Mountains, Tanzania. Tanzania Journal of Health Research. 2014;**16**(3):1-14

[52] Lushoto District Council. Lushoto District Profile [online]. 2016. Available at: http://ded-lushoto.blogspot.com/p/district-profile.html [Accessed: 05-03-2016]

[53] Ward JV, Malard F, Tockner K. Landscape ecology: A framework for integrating pattern and process in river corridors. Landscape Ecology. 2001;**17**:35-45

[54] Gorman OT, Kar JR. Habitat structure and stream fish communities. Ecology. 1978;**59**:507-515

[55] Hastie LC, Boon PJ, Young MR. Physical microhabitat requirements of freshwater pearl mussels, *Margaritifera margaritifera* (L). Hydrobiologia. 2000;**429**:59-71

[56] Chen Y, Yang X, Yang Q, Li D, Long W, Luo W. Factors affecting the distribution pattern of wild plants with extremely small populations in Hainan Island, China. PLoS One. 2014;**9**(5):e97751

[57] Trigas P, Panitsa M, Tsiftsis S. Elevational gradient of vascular plant species richness and endemism in Crete: The effect of post-isolation mountain uplift on a continental island system. PLoS One. 2013;**8**(3):e59425

[58] Meliyo JL, Mansoor HA, Masuki KFG, Mowo JG, German L, Shemdoe RS. Socio-Economic and Biophysical Characteristics of Baga Watershed in Lushoto District Tanzania: A Step towards Integrated Watershed Management in the African Highlands. Report No. 1. African Highlands Initiative Lushoto and the Directorate of Research and Development, Ministry of Agriculture and Food Security, Lushoto, Tanga, Tanzania; 2004